Essentials of Ultrasound Imaging

Essentials of Ultrasound Imaging

THOMAS L. SZABO
Department of Biomedical Engineering, Boston University,
MA, United States

PETER KACZKOWSKI
Verasonics, Kirkland, WA, United States

ACADEMIC PRESS
An imprint of Elsevier

ELSEVIER

Academic Press is an imprint of Elsevier
125 London Wall, London EC2Y 5AS, United Kingdom
525 B Street, Suite 1650, San Diego, CA 92101, United States
50 Hampshire Street, 5th Floor, Cambridge, MA 02139, United States
The Boulevard, Langford Lane, Kidlington, Oxford OX5 1GB, United Kingdom

MATLAB® is a trademark of The MathWorks, Inc. and is used with permission. The MathWorks does not warrant the accuracy of the text or exercises in this book. This book's use or discussion of MATLAB® software or related products does not constitute endorsement or sponsorship by The MathWorks of a particular pedagogical approach or particular use of the MATLAB® software.

Notices
Knowledge and best practice in this field are constantly changing. As new research and experience broaden our understanding, changes in research methods, professional practices, or medical treatment may become necessary.

Practitioners and researchers must always rely on their own experience and knowledge in evaluating and using any information, methods, compounds, or experiments described herein. In using such information or methods they should be mindful of their own safety and the safety of others, including parties for whom they have a professional responsibility.

To the fullest extent of the law, neither the Publisher nor the authors, contributors, or editors, assume any liability for any injury and/or damage to persons or property as a matter of products liability, negligence or otherwise, or from any use or operation of any methods, products, instructions, or ideas contained in the material herein.

ISBN: 978-0-323-95371-9

For Information on all Academic Press publications
visit our website at https://www.elsevier.com/books-and-journals

Publisher: Mara Conner
Acquisitions Editor: Tim Pitts
Editorial Project Manager: Tessa Kathryn
Production Project Manager: Sujithkumar Chandran
Cover Designer: Greg Harris

Typeset by MPS Limited, Chennai, India

Working together
to grow libraries in
developing countries

www.elsevier.com • www.bookaid.org

Contents

For additional information on the topics covered in the book, visit the
companion site: https://www.elsevier.com/books-and-journals/book-companion/9780323953719

Preface

The purpose of Essentials of Ultrasound Imaging is to introduce basic principles of ultrasound imaging to a wider audience including those with advanced scientific training, perhaps in other disciplines, and those with less mathematical and physics backgrounds. This book offers a grounding and explanation of the basic physics and signal processing underlying ultrasound imaging. With an appreciation of the possibilities and physical constraints of ultrasound, readers will be better prepared to understand advanced topics and, hopefully, be encouraged to apply these ideas to research opportunities. The overall presentation of material is unique as it was originally conceived to offer experiential learning opportunities which complement the main text: software simulators and laboratory experiments. This book is a stand-alone text because it offers detailed explanations and basic equations supplemented by numerous figures using examples from the simulators and data and results from the Verasonics® Vantage™ Research Ultrasound System. Nevertheless, there are many instructive advantages to combining this book with the simulators in particular and also with the ultrasound laboratories from the companion curriculum.

A comprehensive set of 27 ultrasound simulators provides an engaging experience for learning about essential physical mechanisms. Each simulator embodies a physical concept and its underlying equations in an attempt to build intuition where sophisticated mathematical analysis or extensive numerical computation was previously required. The simulators have a graphical user interface which typically provides several input controls (sliders and knobs) representing input variables of an equation (or set of equations) in the text. The resulting output is designed to be displayed in near real time in graphical form on laptop computers. In many cases, quantitative information is available to the user such that the simulators serve as virtual laboratories. In some cases, simulators provide tens of thousands or more combinations and offer a wide range of experimentation, and for instructors, many options for teaching and homework problems. For more advanced students, the simulators, combined with the equations in the text, facilitate explorations of parametric relationships and functional intervariable dependencies offering fundamental insights. An additional advantage of the simulators is that the concepts are made accessible to those from other disciplines who may not have had a strong background in wave physics and signal processing. Examples from the simulators are shown in nearly every chapter.

Originally, these simulators were to be embedded in the book itself. Because of the difficulties involved in simultaneously publishing a book and maintaining software, a more practical solution was to separate the two. These simulators were programmed in MATLAB$^{\circledR}$, a high-level scientific programming language, and converted to executable form to run on Apple$^{\circledR}$ or Windows$^{\circledR}$-based personal computers without a Mathworks license required. The simulator package is maintained and available from Verasonics (https://verasonics.com), the company that manufactures the Vantage research ultrasound system used in the labs and offers the companion curriculum.

The Essentials of Ultrasound Imaging Curriculum combines the book and simulators with lecture slides, hands-on experiments and lab kits, and accompanying control and data acquisition scripts to be used with a Vantage system. The Vantage system interface also uses MATLAB, and the scripts are integrated with additional MATLAB programs that process and store results that can be further analyzed by students. This all-inclusive approach combines in-depth learning experiences of the ultrasound imaging concepts with their application. Results from some of the laboratories appear at the end of the chapters.

A brief summary of the organization of the book follows.

- Chapter 1 is an introduction to imaging systems and their functional components, types of medical imaging, ultrasound medical imaging modes and modalities, and a functional block diagram of an ultrasound imaging system.
- Chapter 2 explains types of ultrasound waves and their interaction with media, especially at boundaries and in layers.
- Chapter 3 is about signals, their representation in time and frequency, their manipulation for design, and representation by electrical analog functional blocks and absorption.
- Chapter 4 introduces piezoelectric transducers, how they work, their characteristics, their representation as linear systems, and design goals.
- Chapter 5 shows how beams are formed and focused, described, and quantified.
- Chapter 6 introduces continuous wave arrays which can steer and focus 3D beams electronically, ultrasound-induced heating, and plane wave compounding.
- Chapter 7 covers pulsed phased array beamforming, grating lobes, point spread function, dynamic receive focusing, and types of arrays.
- Chapter 8 provides an overall review of how moving ultrasound images are made and processed; scanning methods, frame rate; and image simulation and measurement.
- Chapter 9 surveys the field of Doppler measurement and Doppler flow imaging, and recent developments such as ultrafast and vector Doppler.
- Chapter 10 opens with a perspective on the growth and relationship among diverse ultrasound fields of study and their applications and provides brief introductions to a range of advanced topics in ultrasound.

Acknowledgments

In reimagining how to present ultrasound in a new way, I am mindful of those I had the privilege of learning from and who are no longer with us: David Blackstock, Wesley Nyborg, Jack Reid, and Marvin Ziskin. I am grateful to them for more than science: lessons of kindness, curiosity, scholarship, and trust (TLS, Thomas L. Szabo).

We thank those at Verasonics who believed in this project for their long-term enthusiasm and support in making this project a reality: Ron Daigle, Ed McClenny, Toni Baumann, Stacy Douthitt, and most of all, Jon Daigle.

Verasonics made it possible to fund a creative team of graduate students of Boston University working part time for several years to bring the ultrasound simulators to life in MATLAB code. For their ingenuity and perseverance, we thank Bowen Song, Sarah Costrell, Elise DeCarli, Joseph S. Greene, Yu Xiao, and Songhao Li. We are thankful for the contributions of Mingxin Zheng for encoding the first simulator programs to show it was all possible. At Verasonics, we are indebted to Juvenal Ormachea and Yi Cheng for reviewing and editing the simulators and to Ryan Ollos for help in packaging the code for distribution. Thanks are also due to Ted Lynch of C.I.R.S. for our collaborations and the development of the Fathom which worked well for our laboratory demonstrations. We appreciate Matt Bruce's comments on parts of the manuscript.

Key parts of some of the simulators rely on the amazingly versatile and fast FOCUS software created by Professor Robert McGough and his team at Michigan State University. The authors appreciate their collaborations and many interesting discussions with Professor McGough over the years.

I am very grateful to my colleagues at Verasonics who contributed to the development of the book and the companion curriculum, for their skill, efficiency, and overall professionalism in developing laboratory software and the phantom kits, foremost Juvenal Ormachea for the Vantage scripts, and Mike Pinch for the lab phantoms with help from Ba Phan in making the kits more elegant and robust. I am very grateful to Aly Chapman and Evan Mladina for consenting to model for the cover, not once but through several "studio" iterations (PJK, Peter J. Kaczkowski).

We are indebted to our editors Tim Pitts and Tessa Kathryn for gently shepherding the authors into fruitful directions and the Elsevier production team ably led by Sujithkumar Chandran for transforming ideas into a beautiful reality.

Special thanks are due to colleagues at Boston University, especially Professor Paul Barbone for enlightening conversations and to Dean Kenneth Lutchen who enabled me to create ultrasound laboratory modules for thousands of biomedical engineering students over the years and for the support of Professor John White and Matt Barber for this project (TLS). TLS is also grateful for participating in the fascinating ultrasound brain imaging with Professors David Boas and Jianbo Tang.

We acknowledge the help of Dr. Francois Vignon, Sue Benzonelli-Blanchard, and Amy Lex and others at Philips for obtaining permissions to use the fine clinical images that grace the book.

Thanks go to those who gave advanced or less known materials and figures: Dr. Rathan Subramaniam, Rajesh Panda, Pat Rafter, Pengfei Song, Jianbo Tang, University of Washington, N.O.A.A., and Kyle Morrison and Francisco Chavez at Sonic Concepts. In addition, we appreciate those who gave us permission to reproduce their work: Daniel Turnbull, Jorgen Jensen, Victor Humphrey, Andrew Baker, and Stanislav Emelianov.

This book is dedicated to my wonderful, clever, insightful wife Deborah who created the extra spaciousness needed for this long book journey by sacrificing her time and leavening my tasks with good cheer, kindness and understanding. The biggest thank you is to you for being there with your encouragement, advice and companionship! (TLS)

I'm extremely grateful to my wife Sara for all of her patience and support during the many weekends and evenings devoted to the work. I can only think of superlatives in recognizing the long list of virtues it took to persevere with grace and generosity while I was engaged in this dream project. The road to completion was much longer and more challenging than I estimated and promised; I thank her for understanding how much this book meant to me and for encouraging me to do it as well as I could. (PJK)

CHAPTER 1

Introduction

1.1 Overview

1.1.1 Prelude

Ultrasound imaging, the fastest growing type of medical imaging, has also found applications in nondestructive testing, sound navigation and ranging (SONAR), and geophysics. Because of recent advances in electronics and digital processing and computation, ultrasound imaging systems have taken many forms from large hospital systems embedded in surgical procedures and ubiquitous diagnostic systems to pocket ultrasound devices. Innovations are continually opening new opportunities in ultrasound.

There is a need for those entering this field of research and development to learn about the underlying physical principles, signal processing, and systems. This book will be particularly useful for those involved or starting out in ultrasound from different backgrounds and skill levels including graduate students, scientists and engineers from other disciplines, physicians and medical professionals conducting ultrasound research, managers of research groups, and those curious about ultrasound science or who are considering or entering the field. A situation frequently arising in industry is that a company will hire employees from other disciplines, and they need to learn about ultrasound imaging to carry out their work. While the content here is primarily focused on graduate students and engineers in the medical ultrasound industry, the material can inform a wider circle of students, instructors, and professionals as well as those involved in application of ultrasound imaging to new areas. As explained later, those who may not have advanced scientific backgrounds can also benefit from this book because of its unique graduated approach.

Essentials of ultrasound imaging offers a fast-track introduction to the science, physics, and technology of ultrasound imaging systems in 10 chapters. It emphasizes the underlying physics that makes ultrasound imaging possible as well as its practical constraints. The interaction of these physical processes with the signal processing and system architectures necessary for image formation are explained in detail. Presentation of the material is unique in two revolutionary ways. First, principles are revealed by examples from software simulation programs which allow students to engage with the concepts with minimal mathematical background. Second, concepts are illustrated with actual data and examples using a Verasonics Vantage Research Ultrasound System.

Essentials of Ultrasound Imaging
DOI: https://doi.org/10.1016/B978-0-323-95371-9.00002-3

The format of the material accommodates different types of readers on four levels. On the first level, the book is a standalone independent source of new introductory material which is drawn from examples using simulation programs which will be explained later in this chapter. On the second level, more advanced material is presented in each chapter in a graduated way including equations for the simulation programs and referrals to more specific detailed explanations available in *Diagnostic Ultrasound Imaging: Inside Out* (Szabo, 2014a). For the third level, names of specific simulators in the text lead to 25 different interactive simulation programs, which readers may access for a nominal fee through Verasonics' website (Verasonics, 2023). Most of the simulators provide quantitative outputs so they can serve as virtual laboratories for homework problems. At the fourth level, the book may be used in combination with the Verasonics "Essentials of Ultrasound Curriculum," which includes lectures, homework exercises using the simulation programs, and hands-on labs using a Verasonics Vantage Research Ultrasound System. More information can be found in Sections 1.8.3 and 1.9.1. Part of this material was well received in an abbreviated format as a 4-hour short course at three IEEE International Ultrasonics Symposia.

1.1.2 In this chapter you will learn

The primary goal of most imaging systems is to make visible the internal structure of opaque material and bodies. How this is done using electromagnetic and acoustic waves is explained in depth. You will be introduced to your own advanced mobile adaptive imaging system based on your eye—brain visual processor. The role of simulators and the first simulators are presented. Medical imaging systems utilizing different physics are compared. How ultrasound images materials noninvasively and nondestructively is revealed. The main types of imaging modes employed to depict three-dimensional (3D) objects are introduced.

1.2 Waves

Waves are disturbances which propagate in a material (gas, liquid, or solid) without changing it. A classic type of wave shape, the sine wave moving in time at a fixed location z, is shown in Fig. 1.1. One cycle of length $T = 0.1$ μs and amplitude $A = 5$ is shown. Because this cycle is a snapshot of an unending sequence of identical cycles, it has a frequency f_0 given by $f_0 = 1/T$. This wave moves with a speed of sound, c_0, and an equation describing this simple wave W having an amplitude A is

$$W = A\ \sin\left[2\pi f_0\left(t - z/c_0\right)\right] \tag{1.1}$$

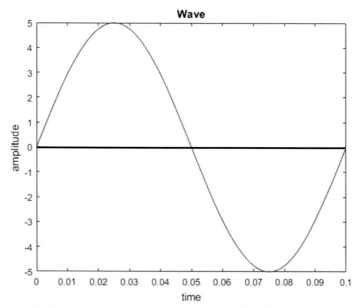

Figure 1.1 One cycle of a propagating sine wave at a fixed value of z. Time scale in microseconds.

This equation shows that a wave traveling a distance z incurs a delay, $t = z/c_0$. For a time scale in microseconds and a speed of sound, $c_0 = 1.5$ mm/μs, the delay is at the center corresponding to $z = 0.075$ mm. A sinusoidal wave has a wavelength, $\lambda = c_0/f_0$ which shows that the higher the frequency, the shorter the wavelength. What are f_0 and λ? This plot displays a cycle of what is called a "radio frequency" (rf) wave which is a general term used to describe a high-frequency wave.

The primary goal of imaging in this book is to reveal invisible structures hidden in opaque materials. By visible we usually mean something seen; but in the context of our discussion, imaging is a picture of hidden structures or features obtained by a process which involves an imaging system. The imaging system acquires data about an object and translates it into an image that we can see. Later in this chapter, we will compare different types of imaging systems which provide very different images of the same object because different physical processes are involved in acquiring the data.

For now, we can stick to ordinary vision for describing how waves interact with different materials. A transparent material, such as glass, allows us to see through it, but an opaque material completely blocks waves from getting through by totally scattering or reflecting them. An intermediate situation is which some waves are scattered; others are absorbed, and the rest are let through. For ultrasound waves in the body, similar processes of scattering, reflection, and through transmission are at work.

1.3 Your very own imaging system

1.3.1 Electromagnetic spectrum

The waves we are most familiar with are electromagnetic. Fig. 1.2 is an illustration of characteristics of the electromagnetic frequencies or spectra. The visible spectrum extends from red at 430 THz (with a red wavelength of 700 nm) to blue at 750 THz (400 nm). Later, imaging at other frequencies will be described and compared.

The most remarkable imaging system is the one to which we have immediate access: our personal eye–brain visual system. Though we usually take this system for granted, it has all the basic elements of a complete imaging system in addition to being portable and adaptive. Before an examination of its components, the overall process, in terms of waves, is illustrated in Fig. 1.3. Here waves from a broadband transmitter, the sun, are sent to an absorbing target which reflects a certain range of colors that are then imaged by our stereoscopic imaging system. This process will be the basis of the first simulator described in the next section.

1.3.2 Digital camera imaging system

First our imaging system will be compared to one with which we may be familiar: a digital camera depicted in Fig. 1.4. The major components are the lens, aperture (controlled by an iris diaphragm), and digital photoelectric array of m by n elements. The

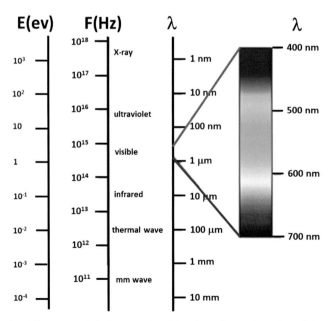

Figure 1.2 Energies (electron volts), frequencies (Hertz) and wavelengths of the electromagnetic spectrum. The visible spectrum is shown as colorbar.

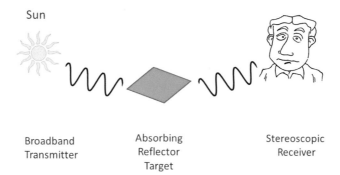

Figure 1.3 Basic components of the human visual imaging system.

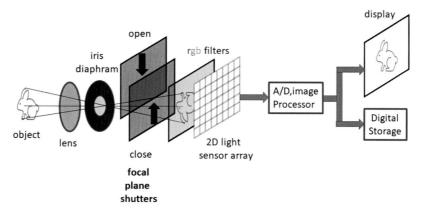

Figure 1.4 Digital camera imaging system consisting of a lens, a light adjusting iris, two focal plane shutters which adjust the time of exposure, a mosaic of red, green, and blue (rgb) filters which convert individual elements in the $m \times n$ light sensor array to detect colors, analog to digital (A/D) converters and an image processor unit which creates a displayed image and stores the image data in as selected format.

lens is chosen to focus the object onto the image plane which is composed of a 2D color mosaic filter to transform individual elements in the array to sense either red, blue, or green colors. The array then converts incoming light into electrical signals that are assembled into an m by n matrix of interpolated color values which are then displayed and stored digitally in a selected format. Present high end digital cameras have about 50 million elements in their arrays. Additional features include exposure control. Since the array has a dynamic range or range of light sensitivity, the amount of incident light for an exposure is ideally selected to be at a midrange level to utilize most of the available range. Exposure adjustment comes from changing the shutter speed and altering the size of the aperture. The aperture adjusts the amount of light passing through its iris diaphragm or "F-stop." The length of exposure is controlled by two focal plane shutters: one which opens to let light pass through and a second shutter which closes, stopping the exposure.

"Automatic focusing" can be achieved by either active or passive means. One of the first active ranging methods was employed on Polaroid Sonar One Step and some SX-70 cameras; they utilized an ultrasound pulse which determined the focusing distance based on the speed of sound in air and the measured roundtrip delay time: $z = tc_0/2$. This approach was superseded by an infrared sensor and later, a passive image analysis method.

1.3.3 Our analog imaging system

Our analog eye—brain visual system is far more complicated than a digital camera, yet some of its features are recognizable. The irises shown in Fig. 1.5 adaptively respond to the amount of incoming light and change the diameter of the pupils, the apertures of the eyes. The focal length of the lens is also adaptive through reshaping by ciliary muscles. Compared to a camera, the eye is much faster and more precise. Because of our stereoscopic vision, the eyes are synchronized and their 3D focusing movement converges on the same depth and horizontal and vertical position.

Images are focused on the retina, a 3D (slightly curved) array of photosensors, the rods and cones. There are about 250 million of these sensors (an order of magnitude more than current digital cameras) which convert light into electrical signals; they are adaptive: cones allow us to see in bright sunlight and rods in starlight (daylight or night vision) to sense a ratio of light intensity levels of 1000 or 30 dB (a logarithmic scale = $10\,LOG_{10}$ [level/reference]). Cones are separated into groups having different color photo sensitivities, each one acting as an individual spectral filter peaking at the electromagnetic light frequency corresponding to either red, green, or blue colors. Six sets of muscles move each eyeball in a coordinated way to change the direction of the vision in the outer landscape; the head can also be moved to increase angular range.

The outputs from these sensors are hardwired into two optical nerve bundles to several parts of the brain including the visual cortex for a considerable amount of image processing. For binocular vision, inputs from both eyes are combined to create high-resolution depth perception. Where is the display located? Complementary and redundant information are seamlessly processed so that the different locations where the optical nerves exit in each eye, blind spots in the retina, are automatically compensated for. The brain also has a "nose filter" which removes the nose from ordinary vision. If you close each eye, one at a time, you will see your nose. Simultaneous with vision is pattern recognition, our ability to recognize people and objects by their features and textures (Fig. 1.5). The eye—brain system also detects movement and can track objects in motion corresponding to about 30 frames per second. Your system is portable by crawling, walking, or running. Finally, your system was delivered free at birth.

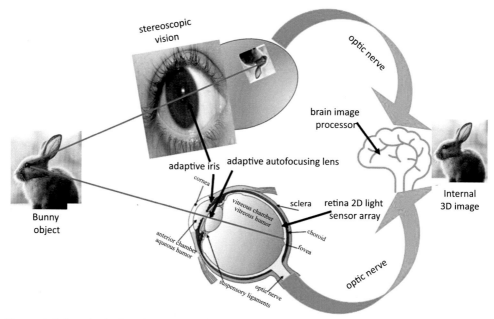

Figure 1.5 Eye–brain imaging system showing two eyeballs with adaptive lenses converged on rabbit object, the two retinal image acquisition two-dimensional sensor arrays, brain image processor producing final three-dimensional image. *Eyes courtesy of Epicstessie. (n.d.).* Own work, CC BY-SA 3.0. *https://commons.wikimedia.org/w/index.php?curid = 16442072; and Fischer, H. (n.d.). Eye: By art-work Holly Fischer. http://open.umich.edu/education/med/resources/second-look-series/materials—Eye* Slide 3, CC BY 3.0. *https://commons.wikimedia.org/w/index.php?curid = 24367145, Wikipedia.*

1.4 Simulators

1.4.1 Introduction to simulators

Most of the concepts described in this book will be introduced through digital simulators. What are they? They are models of physical, signal, or imaging processes. They consist of a single equation, or a series of equations sequenced or coupled together. For example, in a typical engineering homework problem, a student identifies the input variables and boundary and initial conditions and solves the problem by finding the desired output as a function of the input variables. This solution offers an insight of how a concept works under a set of specific conditions. In contrast, computer-based simulators as they are used in this book follow the approach illustrated in Fig. 1.6. A set of input variables $[x_1, x_2, \ldots, x_n]$ is available to the user and accessible through several control options such as sliders and knobs as presented on a graphical user interface (GUI). Similarly, one or more output variables $[y_1, y_2, \ldots, y_m]$ can be selected and displayed using one or more options. In other words, the simulators are the functional equivalents of equations, or series of mathematical operations calculated

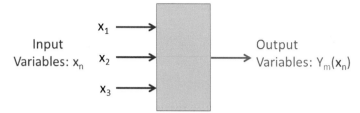

Figure 1.6 Simulator variables.

in near realtime numerically. However, unlike the mathematical homework problems, they provide a greater view of the concept in operation because, typically, there are tens of thousands of combinations of input variables. It is often possible to reveal quickly which input variables have the greatest effect on the output. By the addition of quantification, simulators become virtual laboratories, so that specific numerical relationships among variables can be delineated. The student can spend more time in understanding the concept in a broader set of contexts rather than spending time on numerical computation of a specific circumstance. The concepts can also be under-stood by those who may not have had the benefits of advanced mathematical training. Nevertheless, this book includes many of the equations that underlie the simulators so that the advanced student may benefit from insights the equations can provide.

1.4.2 First example of a simulator

When the essential features of the eye—brain system reduce to a simpler model, the key parts are a broadband transmitter, an absorbing and reflective target, and a receiver. The steps in this spectral approach are shown in Fig. 1.7 in which the frequency spectra associated with the three parts are multiplied together to give an output spectrum. A final step, representing perception, is to translate this output back into a known color patch or image.

The Imaging Systems Simulator can be broken down into three major groups: input and output variables, and display method. In this case, the input variables are transmitter (sun or moon), target spectrum, and eye adaption (receiver) response: day or night. The output variables are the overall imaging system frequency response and the perceived color. For this simulator, there are no display options since there is only one type of display offered. This overall process can be represented by an equation:

$$\text{Output}(f) = G\{[\text{sun}(f) \text{ or moon}(f)] \times \text{target}(f) \times [\text{day}(f) \text{ or night}(f)]\} \qquad (1.2)$$

The output color is determined numerically from a lookup table which converts the output spectrum into a perceived color, $G\{f\}$.

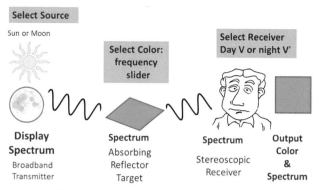

Figure 1.7 Imaging Systems Simulator input and output diagram showing variable selection options.

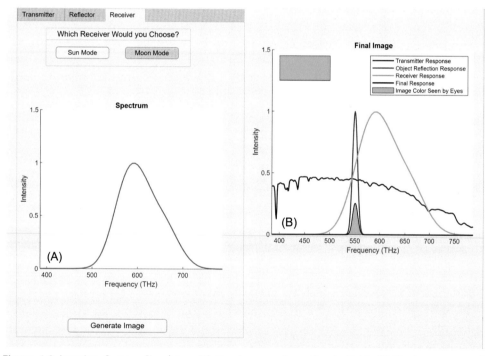

Figure 1.8 Imaging System Simulator. (A) Receiver spectrum for daylight. (B) Final output graph and color for daylight conditions.

An illustration of a calculation from Eq. (1.2) is shown in Fig. 1.8B. The blue curve is the measured solar spectrum which is multiplied by narrow red target curve centered at 550 THz and then by the daylight response of the eyes in green. The final output response is given by the black curve filled in by the perceived color which is shown also as a patch in the upper left corner.

The Imaging System Simulator GUI, programmed in MATLAB, is shown in Fig. 1.8. On the left, the input spectra are selected, in turn, by the tabs at the upper left. As each of the three spectra is selected by a tab or slider below, its shape is plotted below. As shown as the last step, the receiver spectrum is displayed for a sun or day eye response. To the right of the GUI, the overall output spectrum is plotted as a black curve which is the product of the three individual factors: blue curve for the sun transmitter, red curve for the object (centered at 550 THz), and green for the receiver (moon). The color associated with the peak of the output spectrum is shown at the top left corner of the output plot. In the case of moonlight, if the same target frequency was chosen, moonlight conditions and night vision considerably reduce the intensity of the response.

The detailed operations in Eq. (1.2) are executed as follows. A table of numerical values constitutes the daylight transmitted spectrum. The lunar spectrum is also tabulated numerically. The three red, green, and blue (RGB) response curves are consolidated into a composite daytime response spectrum; similarly, a low light level response is created from measured data, The selected target spectrum is represented by a narrowband Gaussian shaped filter. Finally, the output color is determined from the output response curve and a look-up table which relates the peak of the output to a corresponding color and intensity level. Even for this deceptively simple equation, Eq. (1.2), a considerable amount of tedious numerical processing is involved which can be handled easily in real time by a computer.

1.5 Imaging up and down the electromagnetic spectrum

1.5.1 Imaging scorecard

The science of imaging also applies to ways of seeing things that we cannot do with our normal vision. Our vision is restricted to the electromagnetic frequencies our eyes perceive. From sensors that work at other frequencies, other features of surfaces and subsurfaces can be revealed which are normally invisible to us.

In order to compare the features of different imaging systems, an imaging scorecard, outlined in Table 1.1, will be applied first to our vision system. The first question might be: is the system active or passive? Because we do not have a transmitter, unique to the imaging system, under our direct control, our system is passive. Second, what are we imaging? In this case, we are mainly seeing the scattering, reflectivity, and absorption of surfaces and the relative transparency/opaqueness of objects. Third what is the propagation medium? Air, which gives us the speed of light as the velocity in the medium, $c = 3 \times 10^8$ m/s. Fourth, what is the frequency range of vision? From Fig. 1.2, the frequency range is 430−750 THz.

Fifth, what is the object to wavelength ratio? If a is the average object diameter, then this ratio is expressed as a/λ. From questions 2 and 3 or Fig. 1.2, the wavelength

Table 1.1 Imaging scorecard.

Modality	Ultrasound	Vision	IR	mm	X-ray, CT	MRI
1. Active or passive	Active	Passive	Passive	Active	Active	Active
2. What is imaged	Mechanical properties	Surfaces, colors	Surfaces, temperature	Surfaces, reflections	Mean absorption	Biochemistry $(T_1, T_2,$ and PD)
3. Medium	Tissue	Air	Air	Air	Tissue	Tissue
4. Frequency	1–20 MHz	430–750 THz	210–330 THz	0.1 THz	3×10^{16}–3×10^{20} Hz	10's MHz
5. a/λ	10–80	>100	>100	10–80	>1,000,000	N/A
6. Contrast	Reflection	Color, intensity	Temperature	Reflection	Absorption	Relaxation time
7. Energy	Intensity, peak pressure	2.5 eV	1.1 eV	4.1×10^{-4} eV	124 eV–1.24×10^6 eV	$<8.3 \times 10^{-8}$ eV
8. Spatial resolution	Varies: 0.15–3 mm	1/60 degrees	1 mm	3 mm	1 mm	1 mm
9. Range	Varies: 3–25 cm	3 km	100 m	10 m	2 m	2 m
10. Cost	$	Free	$	$$	$$$	$$$
11. Portability	Yes	Yes	Yes	Fixed	Fixed	Fixed
12. Mode	2D and 3D	3D	2D	2D	2D and 3D	3D

range is 700−400 nm. What is one of the smallest objects we can see? Taking a as the diameter of a human hair, about 75 μm, the worst case ratio is $a/\lambda = 75/0.7 = 107$. For a ratio of 10 or more, scattering falls in the specular range or follows the laws of geometric optics, ratios less than this fall in the next lower category where wave theory applies, as explained later. Put simply, objects appear in their normal size and appear smaller or larger depending how far we are from them. This result may seem ordinary or obvious to us, but soon we will be discussing some very strange imaging systems in which these laws no longer hold.

Sixth, what determines contrast in the system? For vision we distinguish objects mainly by color as quantified by the spectral colorbar shown in the right side of Fig. 1.2. We are also responding to different levels of intensity. For example, when we view a black and white (monochrome) picture, a colorbar ranges from black through shades of gray to white in a range of 1000:1 or about 2^{10} or 30 dB.

Seventh, what is the energy of the transmitter? Since light can be viewed either as a wave or a quanta, energy can be expressed as $E = hf$, where h is Planck's constant, 6.63×10^{-34} Joule-second.

More often energy is described relative to an electron charge, e, or in units of electron volts, $E = hf/e = 4.133 \times 10^{-15}f$ (electron volts [eV]). For the visible range, E is about 2.5 eV and is considered a safe level of radiation.

Eighth, what is the resolution of the imaging system? Resolution is defined as the minimum distance between the two smallest objects that can be seen as being separate. Ultimately for a geometric optical system, resolution is determined by the size of the sensors; here, the image is projected on a cone of the retina (about 5 μm in diameter). The measured resolution of the eye for 20/20 vision is 1/60 of a degree.

Ninth, what is the range? For vision, this is determined by the air quality or scattering. We can see stars, and on the ground, objects can be distinguished 3 km away if the air is clear of fog or pollutants.

Tenth, what is the cost of our system? It is free. Eleventh, how portable is the system? Our vision system is totally portable as it goes with us everywhere.

1.5.2 Down the imaging electromagnetic spectrum

Infrared cameras, working at frequencies just below our vision range, can make heated objects visible even in complete darkness. Besides night vision military and security applications, others are detecting thermal leaks in building insulation, seeing through smoke for firefighting and detecting fevers in travelers in airports. Shown in Fig. 1.9 is an image of one of the authors taken at lower frequencies with an infrared camera.

The imaging scorecard can be applied to infrared imaging systems. An infrared camera is usually a passive system, working in air, to image the thermal radiation of objects or people. The frequencies just below our vision range, 210−330 THz, or wavelengths

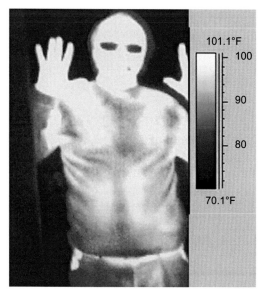

Figure 1.9 Infrared image with temperature scale.

of 9—14 μm. This roughly corresponds to a reduction of about a factor of 10 in resolution from that of the visible spectrum. To obtain approximately the same a/λ ratio as optics, a is about 1 mm; therefore, geometric optical laws still apply but the images are blurrier. Contrast is in terms of the temperature differential between the object and background, and the colorbar is scaled in temperature (degree Fahrenheit) and uses an artificial color mapping scheme as shown in Fig. 1.9. The energy level is about 1.1 eV, half that of visual spectrum. Handheld cameras and smartphone accessories are available for infrared imaging, keeping the cost down; night vision, military, and security cameras employ more expensive smaller sensors. Because camera-like arrays are used, the images are 2D.

The next range of imaging devices fall in the millimeter (mm) range. While mm wavelength imaging is also applied to radio astronomy, our discussion is based on body scanners used for airport security. The frequency often used is 100 GHz or 0.1 THz resulting in a wavelength of 3 mm and a resolution of about 6 mm. An active system employing two vertical stacked arrays swung simultaneously in cylindrical arcs in front of and behind a person operate in a transmit/ receive mode in air. Because cloth is transparent at this frequency, skin provides most of the reflectivity so that the resulting 3D surface image is of a naked body. Controversy over privacy resulted in a outlined display of the naked body on which suspicious items are superimposed. Contrast is derived from the reflectivity of hidden objects relative to that of the skin such as metal guns, knives, plastic bombs, and water bottles. For these objects the a/λ ratio is less than 10, so focusing is needed to achieve the necessary resolution, and images are presented in their correct geometric proportions. The energy is 4.1 × 10^{-4} eV, so this type of imaging is relatively safe. The range is 1−10 m, and the systems are fixed in position and are expensive.

1.5.3 Up the imaging electromagnetic spectrum

High above the visible spectrum shown in Fig. 1.2 is the X-ray band covering projection and computed tomography (CT) imaging. The frequencies involved range from 30 petaHertz (10^{15} Hz) to 300 exaHertz (10^{18} Hz) and wavelengths from 10 nm (10^{-9} m) to 1 pm (10^{-12} m). The energy far exceeds that of the visible spectrum by orders of magnitude, from 124 to 1.24 × 10^6 eV, resulting in ionizing radiation that must be used with caution. The wavelengths involved are so small that the a/λ ratio is in the millions, meaning that geometric optics holds. For X-rays, the medium is the message: the passage of these waves through the body or material is what is imaged. Because the high frequencies involved, the X-ray waves undergo absorption and sometimes, complete scattering. Different materials have various absorption coefficients, and these are codified in a linear scale in terms of Hounsfield units. This color-bar scale is the measure of contrast in a CT image and provides a way to distinguish among different tissues and fluids. Black (−1000) is for air and gray values on the scale become brighter for more absorbing or denser tissues moving through the reference fluid water (0) to full white for bone (+1000).

For active projection or digital X-ray imaging, a source sends X-rays through a body or material and absorption is accumulated along the straight-line paths to form the image. 3D structures of the body are superimposed as a 2D projection onto film or a digital sensor array. The depth location of structures is lost as all this information is compressed into one image plane. Spatial resolution is not determined by wavelength but by focal spot size of the X-ray tube and scatter from tissue. The state of the art is about 1 mm. Projection X-ray imaging requires patients not to move during exposure. Because these are through transmission images, parts of the body that can be imaged are limited to those that are accessible on two sides. Most conventional X-ray systems in common use are dedicated systems (fixed in location) even though portable units are commercially available for special applications. Systems tend to be stationary so that safety precautions can be taken more easily. Though exposures are short, X-rays are a form of ionizing radiation, so dosage effects can be cumulative. Extra precautions are needed for sensitive organs such as eyes and for pregnancies.

To recover depth information of structures, CT utilizes a large fixed circular arrangement of X-ray sources and detectors placed around the body and collects transverse image data over a larger angular range. Through a Radon transform algorithm, the data from tissue structures are transformed to their correct geometric placement and assigned the absorption values according to the Hounsfield greyscale colorbar to produce a 2D image. By sliding the circular array of detectors and sensors along the longitudinal axis of the body, a 3D image can be created. The result has over two orders of magnitude more dynamic range than a conventional X-ray, so subtle shades of the attenuation variations through different tissue structures are seen. This image is an accurate geometric map of internal body structures including bones and, tissues and fluids which can be used for diagnosis, for finding fractures, unusual sizes of organs, tumors, and determining different tissue types, amount and location of fat, etc.

So far, we have used a relative atomic or photon energy parameter to compare imaging modalities; however, actual exposure to radiation depends on the intensity as well as the length of time it is applied. The overall dose of a CT scan is much higher than that of a conventional X-ray, but the same safety precautions as those for conventional X-rays apply. CT equipment is large and stationary in order to fit a person inside it, and as a result, it is relatively expensive to operate. Consecutive pictures of a moving heart are now achievable through synchronization to electrocardiogram (ECG) signals. See Fig. 1.10D for a CT image example.

1.5.4 Magnetic resonance imaging

Magnetic resonance imaging (MRI) produces remarkable maps of the body working at atomic levels. Electromagnetic fields are involved but imaging is not obtained through conventional propagation as in the previously described imaging modalities. Instead, atoms are temporarily excited by applying an external magnetic field and then their decays to equilibrium are detected at different spatial locations to form an image. Like a CT scanner, a person is slid into an MRI scanner in which 2D imaging is done in a plane transverse to the longitudinal axis, z, of the body. This image is known as a slice and by moving the MRI apparatus along the longitudinal axis, a 3D image can be formed. This active system includes a large super conducting magnet which has arrangements of gradient coils to turn on and off a magnetic field gradient. Once the original magnetic moment decays to equilibrium, it emits a relaxation signal which circumferentially placed radio frequency coils detect.

When hydrogen (as a component of water, H_2O, comprises 63% of body weight) is placed in a large static magnetic field, the magnetic moment of the atom spins around it like a tiny gyroscope at the Larmor frequency, which is a unique property of the material. Because a magnetic field gradient, $B(z)$ is applied, different Larmor frequencies are excited according to $f_L = \gamma B(z)$ where γ is the Larmor constant, 42.6 MHz/T. This excitation causes a "spin-up" energy shift, $\Delta E = hf_L$. What makes this type of control possible is that the frequencies used are in the MHz range, so that for a frequency of 20 MHz, $\Delta E = 8.3 \times 10^{-8}$ J. What is imaged is the local responses to these excitations as seen through considerable amounts of signal processing of the received radio frequency waveforms. This voltage signal is detected by coils, and two relaxation constants are sensed. The longitudinal magnetization constant, T_1, is more sensitive to the thermal properties of tissue. The transversal magnetization relaxation constant, T_2, is affected by the local field inhomogeneities. These constants are used to discriminate among different types of tissue and for image formation. The resolution is mainly determined by the gradient or shape of the magnetic field, and it is typically 1 mm. Images are calculated by reconstruction algorithms based on the sensed voltages proportional to the relaxation times. Tomographic images of cross-sectional slices of the body are computed; an example is shown in Fig. 1.10C. The imaging process is fast and safe because no ionizing radiation is used. Because the equipment needed to make the images is expensive, exams are costly.

MRI equipment has several degrees of freedom, such as the timing, orientation, and frequency of auxiliary fields; therefore, a high level of skill is necessary to acquire diagnostically useful images. Diagnostic interpretation of images involves both a thorough knowledge of the settings of the system, as well as considerable experience. For more MRI information, see Szabo (2012) for an introduction and Jara (2013) for advanced explanations.

1.5.5 Ultrasound imaging

Ultrasound imaging is based on transmitting and receiving of scattered or reflected mechanical waves propagating through tissue, material, or fluid. The frequencies range from 1 to 20 MHz and the average speed of sound in tissue is $c_0 = 1.5$ mm/μs; therefore, wavelengths range from 1.5 mm to 75 μm . The a/λ ratio is on the order of multiples of 10 so the aperture and scattering fall in the wave or diffractive range which means ultrasound has a spatially variant resolution that depends on the focusing of the active aperture of the transducer, the center frequency and bandwidth of the transducer, and the selected transmit focal depth. A commonly used focal-depth-to-aperture ratio is five, so that the half power beamwidth is approximately two wavelengths at the center frequency. Therefore, the transmit lateral spatial resolution in millimeters is approximately $2c_0/f_c = 2\lambda$, where f_c is center frequency in megahertz. For frequencies ranging from 1 to 20 MHz, lateral resolution corresponds to 3−0.15 mm.

Another major factor in determining resolution is attenuation, which limits penetration. Attenuation steals energy from the ultrasound field as it propagates and, in the process, effectively lowers the center frequency of the remaining signals as another factor that reduces resolution further. Attenuation also increases with higher center frequencies and depth; therefore, penetration decreases correspondingly so that fine resolution is difficult to achieve at deeper depths; the range is about 25 cm at 2.5 MHz and 2.5 cm at 20 MHz. This limitation is offset by specialized probes such as transesophageal (down the throat) and intracardiac (inside the heart) transducers that provide more direct access to regions inside the body.

Otherwise, access to the body is made externally through many possible "acoustic windows," places where a transducer is coupled to the body with a water-based gel. This "footprint," is the area of contact of contact the transducer makes with the body and is typically a few centimeters squared. Except for regions containing bones, air, or gas, which are opaque to imaging transducers, even small windows can be enough to visualize large interior regions which make ultrasound portable and well suited to point of care imaging. Ultrasound beams are focused and scanned electronically in a plane to provide high-resolution 2D images; 3D imaging is available. Ultrasound measures and displays real-time parameters quantitatively through signal processing such as pulsatile blood flow.

Diagnostic ultrasound has been used safely for over six decades worldwide; two ultrasound-induced bioeffects: heating and cavitation are controlled by monitoring and regulating the estimated in situ values of temporal average intensity and peak pressure. Interpretation of images is based on experience though efforts are underway to analyze images through pattern recognition and artificial intelligence (Szabo, 2014b). An example ultrasound image is presented in Fig 1.10F.

1.5.6 Imaging modalities compared

Now that the imaging scorecard is completed as given in Table 1.1, it is apparent that medical imaging is composed of different representations of the body. Each imaging modality presents unique features and levels of abstraction as apparent in Fig. 1.10 in which each imaging modality reveals a different physical characteristic of a cancer. For positron emission tomography (PET) imaging, the drug F-fluorodeoxyglucose (FDG) is used to enhance tumor imaging due to increased glucose metabolism. PET images of a patient with cancer given the drug FDG are shown in Fig. 1.10. The top left PET images in frontal and transverse views (Fig. 1.10A and B) display the uptake of FDG as a small, dark region (see arrow), signifying hypermetabolic activity and the likelihood of a metastasis; however, the anatomical location of the cancer nodule is ambiguous in these views. The CT view below left (Fig. 1.10D), displays good resolution as well as tissue structures, but not a clear indication of the nodule. The fusion image (Fig. 1.10E), lower middle, a combination of the PET and CT transverse, clearly emphasizes the metastasis with improved location resolution. For comparison, the MRI image (Fig. 1.10C), upper right, of the same view, while providing spatial organization of the organs, is by itself not as informative. Finally, the ultrasound image (Fig. 1.10F), lower right, provides a different view and was useful for a guided biopsy.

All imaging modalities continue to evolve and improve. Image fusion, the combination of different image modalities in complementary ways shows promise as can be inferred from the images shown in Fig. 1.10. Quantitative MRI is

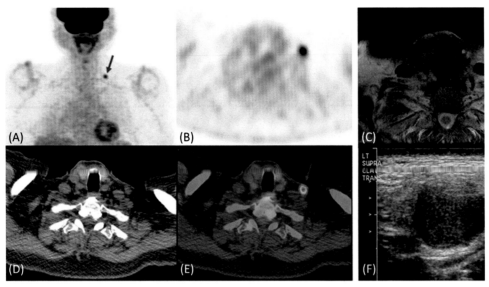

Figure 1.10 Images of different imaging modalities. (A) Upper left: PET frontal; (B) upper middle: PET transverse; (C) upper right: MRI transverse; (D) bottom left: CT transverse; (E) bottom middle: PET/CT transverse fused image; and (F) bottom right: ultrasound. Images of a 74-year-old male patient with nasal cavity esthesioneuroblastoma (a form of nasal cancer) who was restaged on a routine follow-up. PET identified a left supracalvicular hypermetabolic nodal metastases (A, B, D, and E) that was not identified in the MRI (C). An ultrasound-guided biopsy of the node was positive for esthesioneuroblastoma and then surgically resected (F). *Courtesy of Dr. Rathan Subramaniam (2010).*

providing new options to improve contrast for diagnosis (Jara, 2013). Of these imaging methods, ultrasound is the fastest growing imaging modality; it is expanding and adapting its imaging capabilities, contrast, speed, accessibility, portability, and lowering its cost. Beyond diagnostic imaging, ultrasound is being integrated into surgical procedures and performing knifeless surgery in the form of high-intensity focused ultrasound and therapeutics, wound healing, and applications with other types of imaging. The many advantages and technological developments in diagnostic ultrasound are now finding new applications in other fields such as geophysics, nondestructive testing and evaluation, and underwater exploration. An introduction to some of these applications will be provided in Chapter 10.

To take advantage of the potential and to learn limitations of ultrasound, a grounding in the physics of sound propagation and in the means for controlling and directing ultrasound beams from arrays is needed as well as an understanding of the processing of ultrasound waveforms and imaging. A goal of this book is to provide the introductory foundation for further explorations using ultrasound.

1.6 Ultrasound imaging basics

1.6.1 A-line pulse—echo system

Ultrasound waves travel as short mechanical disturbances called "pulses." The average speed of sound in tissues is 1.5 millimeters per microsecond ($c_0 = 1.5$ mm/µs). Typical frequencies in medical ultrasound are in the range of 1—20 megahertz ($f_0 = 1-20$ MHz). Temporal resolution of ultrasound is related to the length of the pulse which is several cycles long or a length of approximately $nT = n/f_0$. For $n = 5$ and a frequency of 5 MHz, the pulse length is 1 µs.

Pulses are sent into a material and if they hit an object, they are reflected. From the time, t, it takes for the sound to reach an object and return and the known speed of sound, the distance to the target can be determined $z = tc_0/2$. Conversely, in an experimental setup, where the distance to a reflector is known or controlled, the speed of sound in the medium can be found from the delay time. The amplitude of the echo can be used to reveal something about the object, its size, or reflectivity as discussed in a later chapter.

An experimental setup for echo ranging is illustrated in Fig. 1.11. A transmitter sends an electrical pulse to a transducer which converts it to a pressure pulse. This pulse hits two objects which each partially reflects and causes pulse echoes shown at the bottom of the figure. Note the pulse echoes have several cycles of radio frequency (shown in red) and an overlaying envelope in black. Shortly after transmission, the

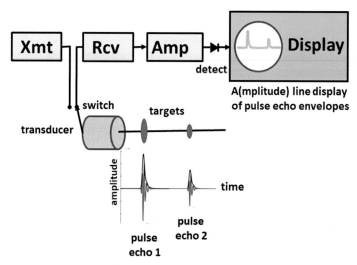

Figure 1.11 Pulse—echo system with an A(mplitude) line display. The transmitter is coupled by a switch to the transducer which converts an electrical pulse to a pressure wave which is reflected by two targets and the resulting pulse echoes are reconverted by the transducer now switched to a receiver to electrical pulses whose envelopes are displayed on an oscilloscope.

returning echoes are sent back through the transducer which converts the pressure pulses into voltage signals and a switch is thrown so that the signals are routed to a receiver where the pulses are cleaned up (noise removed by filtering) and amplified. Finally, the envelopes of the pulses are detected and presented as a function of time on a display like that of an oscilloscope. This presentation is called an A-line mode, where A stands for amplitude.

1.6.2 Ultrasound imaging system

To make a 2D image, another dimension is needed, and it is obtained by a scanning a single A-line in a direction perpendicular to the direction of propagation. Scanning can be done by mechanically moving the position of the transducer in equal increments, or a 1D array can be used. In the latter case, as shown in Fig. 1.12, a switch is thrown sequentially to individual elements of the array. To synchronize and coordinate the scanning operations, a time base is needed.

The object being imaged in this figure is egg-like in shape and has a thin shell which provides the reflecting object to be imaged. Line 1 in the figure at the bottom of the egg, traverses the shell in two places, very similar to the way two objects were imaged by pulse echoes in the A-line example. The imaging process can be viewed as a sequence of A-lines, the positions of which are changed in the upward vertical direction in this figure. Here, line 1 is first and the scanning continues past the top of the egg in line 7. If we were to view the pulse echo envelopes along a timeline from above, we would see blobs at the locations of the pulse echoes in time along the line.

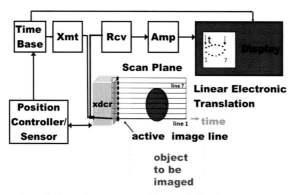

Figure 1.12 B(rightness) mode imaging system with a switched array as a scanning mechanism. A time base orchestrates the position controller which manages the sequential switching of each array element between the transmitter and receiver paths. Each array element sends and receives a pulse echo timeline which is amplified, envelope detected and shown with time in a downward direction on a display. Each line 1−7 intersects the egg and the pulse echoes shown as dark points in the display, trace out the shape of an egg in the image (Szabo, 2014c). *Adapted from Szabo, T. L. (2014c). Overview. In Diagnostic ultrasound imaging: Inside out (2nd ed.). Elsevier.*

In a brightness mode or "B-mode," a series of A-lines are presented side by side with the time axis vertically downward, as illustrated by the display image in Fig. 1.12. Here we see lines 1−7 rearranged with increasing time downward. Black here represents maximum amplitude. The blobs or echoes from each line trace out the shape of the eggshell rotated 90 degrees in its orientation relative to the position of the array and object shown in the figure. Of course, the same image would be obtained had the egg been horizontal and the array positioned above it. Note that image lines are stored so that all the lines can be viewed simultaneously. The lines form a "scan plane" or a 2D image of the center cut plane through the middle of the egg.

1.7 Imaging three-dimensional objects

1.7.1 Imaging modes

As implied by the last section, ultrasound imaging often produces a 2D slice of a 3D object. Fig. 1.13 depicts a 3D teardrop. In the upper left corner of this figure, an ultrasound A-line pierces the teardrop. Note the implied position of the single transducer is at $x = 100$ and $y = 100$ with the depth or time dimension traveling along z from 0 to 200. The A-line display is at the lower right of the figure. For the B-mode, the

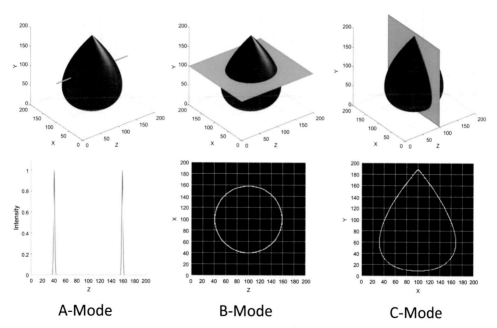

Figure 1.13 A, B, and C imaging modes of a teardrop. Above, the relation of the line or scan plane to the object and below, corresponding A, B, or C mode images.

cut plane at the top middle of this figure is positioned at $y = 100$. In this case, an array is aligned along the scan dimension x, is positioned at a height of $y = 100$, and has an x-length of 200; and the scan extends to a depth of $z = 200$. The corresponding B-mode image is shown in the lower middle plot in figure. At the upper right corner of the figure, an array, positioned in the plane at $z = 0$, extends from $x = 0$ to 200 and scans from $y = 0$ to 200. The scan plane is at the depth $z = 100$ and is a cross-sectional or transverse cut plane through the teardrop. The corresponding image is below at the lower right corner of the figure.

1.7.2 Three-dimensional imaging modes simulator

In general, the 3D object or collection of objects being imaged is unknown, which makes 2D ultrasound imaging interesting. One must infer what is the shape and size of the object from limited information. In medical imaging, the transducer is often moved slightly to obtain a better sense of the 3D nature of the object and our brain creates a mental picture of the object from the additional information.

The 3D Imaging Modes Simulator provides opportunities to play with this limited viewing experience as shown in Fig. 1.14. The images for Fig. 1.13 were created by

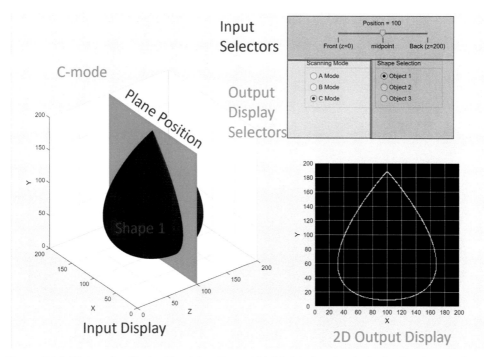

Figure 1.14 3D Imaging Modes Simulator: the red highlighted region indicates input controls and green area, output display choices.

this simulator. On the left side of the GUI is the selected 3D shape with a green scan plane through it. Note the object is placed in a 3D Cartesian coordinate system. In the upper right corner of the GUI, object input variable number 1 is selected, corresponding to the teardrop shaped object on the left side. When the display mode is chosen; C-mode in this example, the other input variable is z position of the $x-y$ scan plane. As the input variable slider above is adjusted, the z position of the scan plane moves through the object, and the image changes correspondingly. At the lower right is the output display plotted on a grid. This image changes as the cut plane or line position or other imaging modes are selected. Different objects are available to view. Note that each imaging configuration implies a transducer or array position and type as explained in the previous paragraph. Because the image is shown on a grid, quantitative information can be extracted from the image. This feature is useful in the lab for an exercise in which the shape of unknown objects are deduced from images and their placement.

1.8 Ultrasound imaging systems

1.8.1 Ultrasound imaging system block diagram

As the functioning of an ultrasound imaging system and processes are described in more detail in coming chapters, both physics and signal processing are involved. The overall combination is represented in Fig. 1.15 with the blocks representing physical processes highlighted by colors. The regions outlined in red correspond to the three basic elements used in the first simulator: a transmitter, object, and receiver. The lenses in our eyes focus images onto the 2D arrays of light sensors, rods and cones, after which the visual data is processed into stereoscopic images by our brain. Ultrasound imaging systems also have adaptable lenses and image processors for forming images. In active mm-wave imaging systems, a linear array of sensors transmits electromagnetic energy and receives echoes from near the skin surface. Ultrasound also employs pulse echoes, but ultrasound penetrates the interior of the body to provide information about hidden internal structures.

Fig. 1.15 provides a preview of the related processes just described. The transducers are the array sensors in this case and translate signals between the acoustic and electrical domains. The electrical beamformers and physical process of diffraction focus the beam and act as acoustic lenses. The ultrasound imaging system creates a 2D map of the variation of reflector—absorber—scatterer values as they change from point to point in the image. The pulse—echo data is organized into a display format during the last stages of receive processing. While this information about imaging systems is informative, it still does not tell us how an ultrasound system is physically realized.

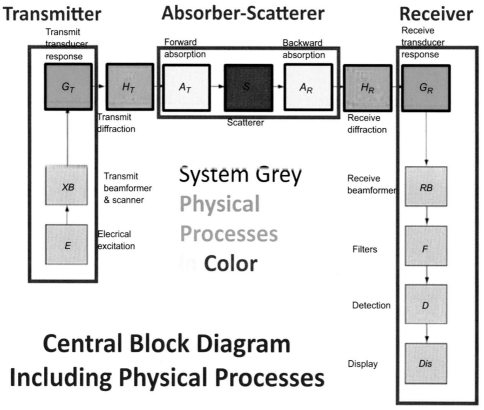

Figure 1.15 Block diagram of an ultrasound imaging system illustrating the components of the pulse—echo system, physical scattering processes that produce backscattered ultrasound energy returning to the transducer, and the signal processing steps required to form an image from the data and display it. *Adapted from Szabo, T. L. (2014c). Overview. In Diagnostic ultrasound imaging: Inside out (2nd ed.). Elsevier.*

1.8.2 Introduction to the Verasonics Vantage Research Ultrasound System

Medical ultrasound systems must provide high-quality sonographic images at frame rates sufficient for real-time clinical visualization. To achieve desired spatial and temporal resolution, many independent signal channels are required, and system designers have relied on dedicated hardware solutions to perform the data processing tasks for achieving the high data rate. The downstream processed information is greatly reduced in size over the raw incoming channel data and is thus easily transferred to the host computer for annotation and display. Fig. 1.16 shows that the data flows continuously through a conventional system starting from the transducer at the left and moves to

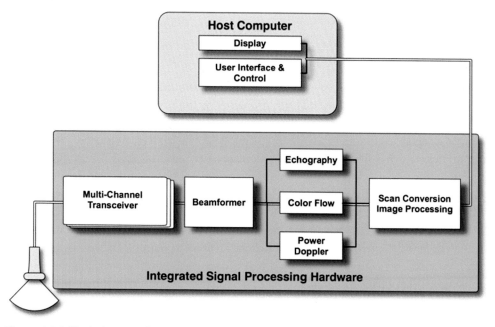

Figure 1.16 Block diagram illustrating the components of a conventional hardware-based medical ultrasound imaging system. The diagram shows that data management and display processing is managed by a computer but that the intensive imaging processing is done in hardware. Pulse–echo data streams in from the transducer on the left and enters electronically addressable hardware for processing the incoming information. The flow is roughly as follows: data is cleaned up and amplified and then digitized and sent to the beamformer for focusing, routed to one of several selected imaging mode processors and finalized into a display format, then sent to the computer for final annotation and display.

the right, with a dedicated processing path for each imaging mode, illustrated as echography (B-mode), and two Doppler imaging modes.

These hardware-based systems are costly to build due to reliance on large, high-density circuit boards with custom integrated logic circuits, coupled with relatively low manufacturing volume. Such systems are also expensive to develop and to upgrade because hardware design cycles are long. Clinical customers do not have access to the latest developments in ultrasound methods and technology because of the long design cycles and the need for regulatory approval for changes to established designs.

Verasonics has been developing software-based ultrasound systems to greatly reduce both manufacturing cost and development time. Innovative hardware and software solutions have permitted shifting most of the tasks conventionally requiring custom hardware (Application Specific Integrated Circuits, or ASICs) to software running on the host computer without compromising clinical performance. Relatively inexpensive general purpose personal computers are now highly capable computational tools, commonly providing large memory sizes, very fast processing, and fast expansion bus communication to external devices. Nevertheless, software-based ultrasound systems are only practical if the standard "flow-through" data processing approach, in which all channel data are processed continuously as they are collected, is replaced by an algorithm which processes only those data that are required to form a given image. This patented technique, called "pixel-oriented processing", greatly reduces the memory bandwidth and processing load requirements of the system (Daigle, 2017). Direct implementation of a conventional flow-through architecture in software would result in an unacceptable computational load, estimated to be at least two orders of magnitude higher, and has likely led other designers to dismiss software-based approaches. The Verasonics architecture is diagramed in Fig. 1.17

The Verasonics hardware "front end" is a programmable data acquisition system with analog transmit/receive circuitry, analog to digital conversion, local memory, as shown by the green block in Fig. 1.17, and a peripheral component interconnect express interface. The system includes a set of transmitters to excite an array to transmit an acoustic beam and in receive mode, it sends a minimally processed data set to the host, retaining maximum flexibility. It is important to emphasize that no receive beamforming is done in hardware, and the raw data rate to the host is indeed very large: the Vantage acquisition system can transfer up to 6.6 GB/s continuously. Once the data has been acquired, the host computer can process it by easily modifiable software (tan block). The resulting flexibility is an

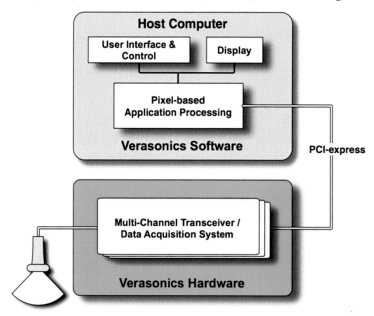

Figure 1.17 Block diagram illustrating the components of the Verasonics ultrasound system, based upon a software-centric architecture in which only the data acquisition and signal conditioning tasks are performed in hardware, and all other image formation and display processing is efficiently performed in software on a general-purpose computer.

advantage to researcher and application engineer alike: it is straightforward to compare different processing schemes under realistic "real-time" conditions. Implementation of new data acquisition sequences and beamforming algorithms for application development and optimization is feasible because the Verasonics hardware performs a relatively simple data acquisition and signal conditioning tasks and does not require customization or redesign for a wide variety of applications.

Once on the host computer, novel beamforming and image reconstruction techniques are used to further reduce the processing load to the minimum required to produce the image of interest. These include the pixel-oriented processing technology outlined below, as well as parallel execution using multiple cores and processor-specific optimization using SIMD programing (Single Instruction Multiple Data vector processing).

The Verasonics system is extremely flexible to program because of the essential simplicity of the data acquisition hardware, and the use of a sequential modular programming structure. The operation of the system is defined as a chain of "events" that define transmit and receive conditions, application-specific image reconstruction procedures, and annotation and user display. Different imaging modalities are straightforwardly specified by different event sequences, all defined in software and saved in a control file. It is comparatively easy to optimize or reorder a sequence for a particular ultrasound mode, and to develop completely new ultrasound modalities than rely on sequences that are impractical to implement in a fixed-hardware environment.

Finally, Verasonics has included a unique capability: real-time simulation of the entire system, using a raw data software simulator based on a medium populated with point reflectors that runs on the system host computer (Windows, Mac OS X, and linux operating systems are supported). This internal simulator is very tightly integrated with the hardware system; indeed, the same sequence control file format used to program hardware execution is used to drive the simulator. The software architecture of the Verasonics system permits exploring the use of different processing code for prerecorded element-level data, making the simulator an invaluable development tool for optimization of processing schemes and evaluation of real-time frame rates. The complete compatibility of event sequence file structure permits instant porting of simulated control sequences to the hardware for experimental evaluation with the hardware and new target media.

In summary, the Verasonics Vantage Research Ultrasound System is an open system, that is, it is programmable in MATLAB by the user to conduct imaging experiments that range from conventional approaches currently used on clinical systems, or on instruments intended for materials evaluation (non-destructive evaluation), to radically new ways of using ultrasound to probe an optically opaque medium. The architecture pairs a high-performance pulse—echo data acquisition hardware system with a computer that performs all imaging tasks in software. The guiding principle of software-based open systems is to permit a single ultrasound developer to conceive, implement and evaluate a new approach very quickly.

1.8.3 Imaging with the Vantage system

A complete ultrasound system based on the Verasonics platform includes the Vantage data acquisition system, a host computer running the Vantage software within a Matrix Laboratory (MATLAB) environment, an ultrasound array transducer, and a test target (phantom). A typical arrangement of the equipment is depicted in Fig. 1.18. Contact with the system is made through display monitor of the host computer (tan block in Fig. 1.17). Here different scripts can be run to acquire and process ultrasound data (shown as A-line radio frequency data at the top of the screen) and produce a final output display shown below on the monitor display. On the left is the script used to process and display the data. For the purposes of the laboratories, specialized scripts and GUIs will be provided for each one. To the left of the computer is the Vantage hardware (green block in Fig. 1.17). Plugged into the hardware unit is the ultrasound transducer which is lying on an ultrasound tissue-mimicking phantom.

The software-based architecture permits the Vantage to be used in a conventional imaging mode or programed by the user to perform novel types of scans. This flexibility was essential in developing new imaging and display modes for the course which seeks to simplify instrument control for the student and create data acquisition and processing flows designed specifically to support the lecture topics.

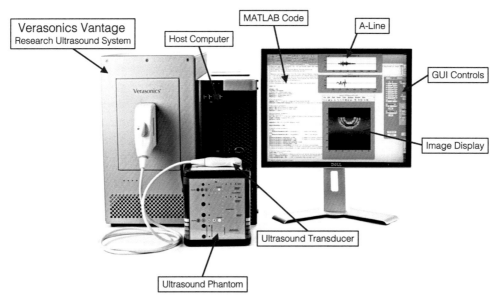

Figure 1.18 Annotated photograph of an ultrasound imaging system using the Verasonics Vantage software and hardware. The Vantage hardware is a multichannel pulse—echo acquisition system only; image formation and postprocessing is done entirely on the host computer.

The simulators and lab exercises are part of an introductory ultrasonics curriculum, The Essentials of Ultrasound Imaging, offered by Verasonics (2023). This course includes the simulators, a manual and exercises, a set of 10 lectures corresponding to the chapters in this book, and laboratory exercises and lecture notes as well as lab scripts, lab lecture notes, and lab phantoms.

In the first lab, the student is introduced to the Vantage system and its components, including a description of the transducer and the imaging phantom. The student is then encouraged to manipulate the transducer and image the phantom, exploring the role of each degree of freedom with which the transducer can be positioned. The student learns about the need for a coupling medium between the target and the transducer face and captures images of 3D objects using various imaging planes. The student also conducts a secret object imaging assignment in which the shape of a test target in an opaque medium must be determined by ultrasound scanning to form an image of a recognizable iconic shape. This procedure reinforces the importance of careful transducer manipulation and provides a different representation of a familiar object; the student is asked to compare the optical (visual and photographic) image to the ultrasound image, noting differences in image quality that will later in the book be tied to the physical processes involved in ultrasonic image formation.

Labs in subsequent chapters are customized to examine various features of the ultrasound imaging system, including the backscattered ultrasound data from the array elements and the beamformed (reconstructed) data under various conditions and backscattering from specific targets. Students will observe and measure the point spread function for several different imaging approaches, as well as the impact of using a variety of excitation signals. Each lab uses Vantage software with its own customized GUI and processing tools, developed specifically for this curriculum.

1.9 Lab 1: Two-dimensional imaging in a three-dimensional world

Laboratory work is integral to the introductory ultrasound imaging course that this book is intended to augment. As mentioned, laboratory exercises are divided into two components: work with the self-contained MATLAB simulator applications to explore concepts introduced in each chapter without relying on manipulation of equations, and practical exercises using the Vantage Research Ultrasound System to give students exposure to many aspects of a real ultrasound imaging system and its use in collecting and processing of ultrasound data. Although students working from the book alone can explore concepts with the GUI-based simulator applications, students in the instructor-led course will be able to conduct experiments with the Vantage system, a transducer, and several phantoms (test targets). Some of these experiments will be described at the end of each chapter, and examples of results will be presented to further explain the chapter content.

1.9.1 Exercises with simulator applications

The simulators were introduced in Section 1.4. This collection of GUI-based programs is available for an additional fee and can be installed on a personal computer using either Windows 10 or Mac OS X. Please see the website (Verasonics, 2023) for further information on computing requirements and compatibility. Though the simulators were developed using MATLAB scientific software, the simulator applications are in an executable form which does not require a MATLAB license. More details are available at Verasonics.com.

All simulator programs are accessible via the control board, a graphical application launcher with the interface as shown in Fig. 1.19. The control board is displayed when the simulator software application is opened on the computer. Each simulator is launched by expanding the directory tree for the appropriate lecture and clicking the name of the simulator.

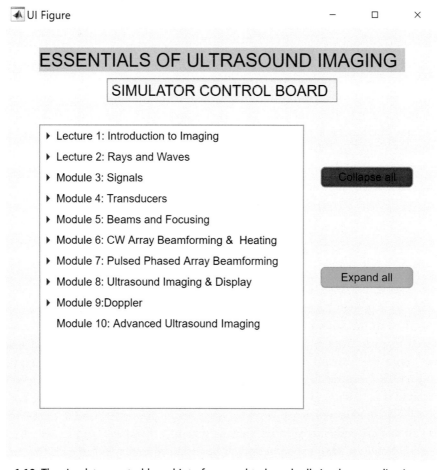

Figure 1.19 The simulator control board interface used to launch all simulator applications.

Each chapter in this book corresponds to a lecture in the course, and the simulators are organized by lecture. In addition, programs are intended to augment the student's understanding of the lecture material, and for exploration during the laboratory period and for homework. Students in the instructor-led labs will share access to a Vantage system, and work with the simulators while waiting their turn.

Detailed information about each simulator program is provided in the simulator manual that comes with the simulator software. For each program, the manual presents a list of the input and output variables and describes output display options as appropriate.

1.9.1.1 Secret Object Simulator

This exercise is intended to help the student visualize a 3D object when collecting data obtained only as 1D or 2D slices through the object. The Secret Object Simulator is based on the 3D Imaging Modes Simulator used earlier in the chapter but hides the object from view so that the student can "scan" various lines or planes through the unseen object and sketch it out to reveal its shape. Our mind is often able to compile the information gathered in this way and interpret the data to produce a 3D model, a remarkable feat, especially in the case of A-mode data! In this laboratory, the student is encouraged to consider questions such as:

- How many data scans does it take to determine the 3D object?
- How finely spaced do the scans need to be, using A-mode and B- or C-mode to properly delineate the 3D shape?

1.9.1.2 Imaging Systems Simulator

This lab is intended to give the student an appreciation for the compound effect of imaging system components and processes on the result. Specifically, by using the Imaging Systems Simulator introduced in Section 1.4.2, the apparent color of the object being examined depends on the combination of color spectrum properties of the light source, the reflecting object, and the receiver leading to the final perceived color. We can compare this approach to our own perception of object color under various kinds of lighting conditions, using our marvelous portable and built-in imaging system, namely our eye—brain system (though we neglect the cognitive processing in our visual system). This simulator is presented again in Fig. 1.20. This figure indicates the sequence of steps needed to produce the output and is an example of the kind of detailed information in the simulator manual.

The student can choose various properties for the illumination and the eye's sensitivity spectra, and the reflectance spectrum of the object to assess the final color and the intensity of the reflected light. Because the result is a product of the spectra, the object will appear dark if any of the spectra are low in magnitude in the object's optical reflectance band. The laboratory provides specific exercises to try.

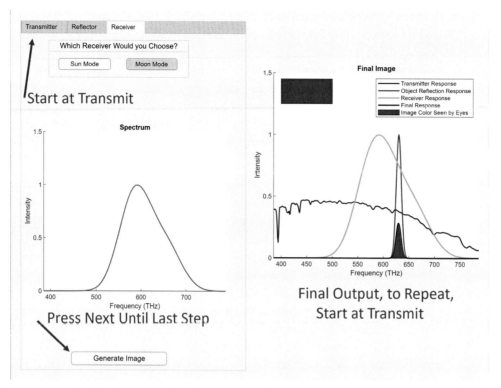

Figure 1.20 The interface to the Imaging System Simulator is presented here. The spectral color properties of the transmitter (the light source), the object, and the receiver (the eye) are chosen sequentially to produce the plot of the relevant spectra and resulting color.

1.9.2 Experiments and exercises with the Vantage system

The goal of this portion of the lab is to familiarize the student with the components of a research ultrasound system, and in so doing, to use the system to perform some simple manipulations of the transducer to image two types of objects: a commercial contrast and resolution phantom, and some common 3D objects that might not be easy to identify using only a 2D imaging plane. The student will observe that physically aiming the transducer to obtain a good view is not easy, and that the ultrasound image is not only a very thin 2D imaging plane but produces an imperfect rendering of the object under investigation; in essence, ultrasound does not perceive the world as our eyes do.

The resolution and contrast imaging phantom used in this course is the Computerized Imaging Reference Systems, Inc. (CIRS) "Fathom," depicted in Fig. 1.21.

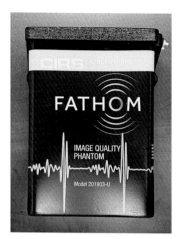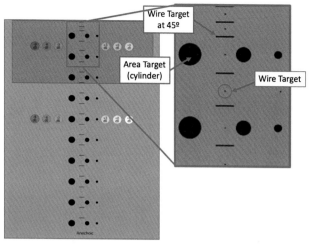

Figure 1.21 The CIRS Fathom phantom (left image) is designed for ultrasound imaging contrast and resolution measurements. The background medium is made of a synthetic rubber with a distribution of tiny particles to create a speckle pattern, filaments to create bright point-like reflectors, and wall-less cylinders with varying degrees of particulate concentrations to provide contrast targets. The target map (right images) indicates the spatial arrangement of the target objects. Only the first 5 cm or so can be imaged with the transducer used in the labs (pink overlay).

The students access the Vantage system through the GUI presented in Fig. 1.22 which contains very few controls. Additional controls will be provided for the labs in subsequent chapters as additional concepts are introduced. With this imaging mode, students can image the Fathom phantom and obtain and capture images such as those shown in Fig. 1.22.

Note that in this figure wire targets and cysts are ideally circles in a transverse view (middle image) but are linear or rectangular in shape in a longitudinal view (right image) similar to the cylinder viewed in the 3D imaging modes simulator.

Students are encouraged to explore the phantom using various transducer orientations, and to become familiar with translation and rotation of the probe, and the importance of the coupling fluid (water). The manipulations and observations solidify the concepts introduced by the imaging mode simulator. Advanced students can collect images onto a removable universal serial bus drive and further process them using a special purpose program to create spatially scaled plots of linear profiles of their choosing through an image. Thus, properties of the ultrasound image can be extracted and plotted, and further processed if the students are able to program in MATLAB. Concepts of dynamic range, signal to noise ratio, 8–bit image scaling, and others can all be illustrated with concrete examples of images obtained by the students themselves.

The second part of the Vantage lab uses a custom tank which permits insertion of different targets for a suite of experiments to be conducted over the duration of the

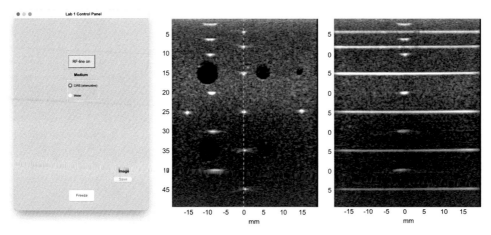

Figure 1.22 The control graphical user interface for the first lab (left image) provides only rudimentary controls to permit imaging the rubber phantom and a "secret object" water phantom. The labs use a linear array with a center frequency of about 7.5 MHz. An ultrasound image of the phantom with bright point targets (middle image) is presented to illustrate how the targets appear when imaged in the conventional view. Rotating the probe by 90 degrees and in the plane indicated by the dashed green line, a longitudinal view of the wire targets is presented (right image). The phantom includes a second set of wire targets, visible in both images, that are oriented at 45 degrees to the main structures for measurement of the elevation focal depth, as discussed in Lab 6.

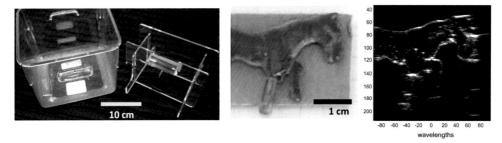

Figure 1.23 Image of the small custom acoustic phantoms (SCAP) tank and insert frame (left) and one of the special target inserts (middle) are presented in this figure. The rubber horse can be made invisible by adding milk to the water coupling liquid, leaving the identification of the target as a challenge to the student. Proper alignment of the transducer leads to a recognizable ultrasound image of the horse (right).

course. When filled with water, the targets can be visually examined, but use of milk or other opaque liquid requires the student to determine the shape of each object using ultrasound alone. Fig. 1.23 shows the tank and one of the inserts with a toy animal horse made of soft rubber and includes an ultrasound image of the horse. Once again, the student must orient the transducer with care because the rubber shape is only 5−6 mm thick, and up to 4 cm in length.

The students are encouraged to compare ultrasound images to those obtained optically, both by using their visual system and by photographing the shape using their mobile phone camera. By downloading the ultrasound images, students can apply additional analysis software to quantify the ultrasound data to measure the size of objects and their features. The lab is intended to be an introduction to the world of ultrasound imaging, and motivation to learn about how ultrasound works.

References

Daigle, R. E. (2017). *Ultrasound imaging system with pixel oriented processing* (US patent 9,649,094 B2).

Jara, H. (2013). *Theory of quantitative magnetic resonance imaging.* Hackensack, NJ: World Scientific Publishing Co., Pte. Ltd.

Szabo, T. L. (2012). Medical imaging. In J. D. Enderle, & J. D. Brozino (Eds.), *Introduction to biomedical engineering* (3rd ed.). New York: Academic Press, Elsevier, Chapter 16.

Szabo, T. L. (2014a). *Diagnostic ultrasound imaging: Inside out* (2nd ed.). Oxford: Elsevier.

Szabo, T. L. (2014b). *Diagnostic ultrasound imaging: Inside out* (2nd ed.). Oxford: Elsevier, Chapter 1.

Szabo, T. L. (2014c). *Diagnostic ultrasound imaging: Inside out* (2nd ed.). Oxford: Elsevier, Chapter 2.

Verasonics. (2023). https://verasonics.com/. Accessed March 1, 2023.

CHAPTER 2

Rays and waves

2.1 Overview

2.1.1 Introduction

One of the interesting conclusions from comparing different imaging systems is that for those which use waves, the ratio of the largest dimension of an object imaged to wavelength is important. For objects much greater than a wavelength, the principles of geometrical optics apply. Under these conditions, waves are assumed to travel in straight lines called "rays." When the objects approach wavelengths in size, diffractive conditions apply and special treatments are needed to account for their seemingly strange behavior and these are treated later in Chapter 5.

In order to build up more realistic descriptions of ultrasound propagation and focusing and imaging, a foundation must be laid first. Certain simplified descriptions of ultrasound behavior are convenient starting points. Later, realistic details will be filled in.

The first simplification, that of an unending sine wave, is introduced in Fig. 1.1 and Eq. (1.1). Of course, in reality, a wave has to start and stop sometime, so this continuous wave is often approximated by numerous cycles, collectively called a "tone burst." Mathematically, a sine wave is defined as

$$\sin(a) = \left(e^{ia} - e^{-ia}\right)/2i. \tag{2.1}$$

There is also a cosine wave which is 90 degrees out of phase with the sine wave and is

$$\cos(a) = \left(e^{ia} + e^{-ia}\right)/2. \tag{2.2}$$

Each exponent is a complex number with a real and an imaginary part, defined as

$$e^{ia} = \cos(a) + i\sin(a), \tag{2.3a}$$

and

$$e^{-ia} = \cos(a) - i\sin(a). \tag{2.3b}$$

In these two equations, $\cos(a)$ is the real part and $\pm i\sin(a)$ is the imaginary part where $i = \sqrt{-1}$ is used to distinguish between the real and imaginary parts. Note "imaginary" has a specific mathematical meaning and exists as a defined term, not as a product of imagination. The parts of the exponent are represented geometrically in

Essentials of Ultrasound Imaging
DOI: https://doi.org/10.1016/B978-0-323-95371-9.00001-1

Fig. 2.1. If we multiply an exponent by an amplitude A, written alternatively as $A \exp(ia)$, it has a magnitude A from

$$\text{magnitude}\left(Ae^{ia}\right) = A\sqrt{\sin^2(a) + \cos^2(a)} = A, \qquad (2.4a)$$

and a phase,

$$\text{phase}\left(Ae^{ia}\right) = \text{arctangent}\left[\sin(ia)/\cos(ia)\right] = a. \qquad (2.4b)$$

2.1.2 Pulse Delay Simulator

If $a = \left[2\pi f_0\left(t - z/c_0\right)\right]$ in Eq. (2.1), we have a sinusoidal shaped wave, $\sin(a)$, moving to the right along the propagation axis, z. Even though in a continuous wave all the individual cycles are identical, the wave still moves to the right. If an insect riding the wave had the uncanny skill of telling these cycles apart, it could hitch a ride on a particular position of selected cycle and move with a speed c_0 which for this example, is 1.5 mm/μs. This motion is illustrated by the Pulse Delay Simulator, using a single cycle of a sine wave, as indicated in the top of Fig. 2.2. Another insect that was going to travel in the opposite direction, that is,

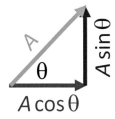

Figure 2.1 Real and imaginary components of A.

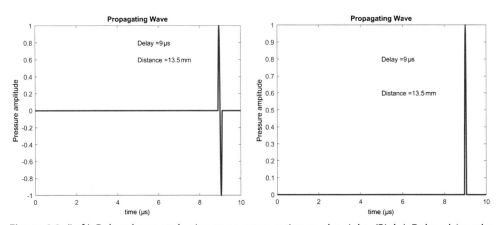

Figure 2.2 (Left) Delayed one cycle sine wave propagating to the right. (Right) Delayed impulse propagating to the right.

along $-z$, would hop a ride on a sine wave with an argument, $a = \left[2\pi f_0\left(t + z/c_0\right)\right]$. This situation will occur for a pulse echo in which a pulse moves to the right, hits a target, and is reflected so that the echo signal moves back along the opposite direction, $-z$.

The second simplification is the use of an extremely short pulse in time, or impulse, represented by the symbol, "δ." A pulse moving along the z axis is described by $\delta\left[2\pi f_0\left(t - z/c_0\right)\right]$. The practical application of an impulse is in situations where the pulse width in time t is much smaller than the temporal resolution of the system in which it is used. This type of wave is shown in the right half of Fig. 2.2.

Using these two extreme forms of waves, we can explore different types of acoustic waves and how they propagate and interact with materials. With the use of simulators, we can examine the shape and direction of waves and the consequences of representing them as rays or waves.

2.1.3 In this chapter you will learn

To understand the basics of wave propagation (Pulse Delay Simulator), some simplifying assumptions can be made. Two extreme kinds of waves are considered: those that are always on—continuous waves, and those which are very short in time—impulse waves. These approximations will facilitate the visualization of three forms of wave geometry: plane, cylindrical, and spherical; these are common simplified shapes of expanding wavefronts. Another useful approximation is to treat waves as either being rays or continuous plane waves. Plane waves are the most often used, with a plane of infinite lateral extent perpendicular to the propagation axis. Rays are well described as vectors with a magnitude and a direction of propagation. When the magnitude of the vector is a wave number $k = 2\pi f/c$ with a direction, it becomes a k-ray convenient for describing waves at boundaries between different materials. By comparing acoustic parameters to analogous electrical circuit variables, models from electrical engineering can be adapted to acoustics for treating waves at boundaries and layers to determine the transfer of pressure or power across the boundary. Interesting things can happen when a wave enters a layer sandwiched between two other materials. The effects are explored in two simulators, one for extremely short impulses reverberating in a layer and another for a continuous wave in layer. Finally, types of elastic wave vibrations, longitudinal and shear, are discussed. Because most of the simulators in this chapter are quantitative, they offer new ways to explore and characterize wave behavior under many conditions.

2.2 Acoustic/electric analogs

Acoustic waves can be thought of as being analogous to well-known electrical parameters. Later, acoustic and electrical parameters will be needed to describe how transducers work, so knowing about both acoustic and electric variables is a good investment. For a voltage V, and a current I, the impedance is defined as $Z = V/I$. In the left side of Fig. 2.3, a sample

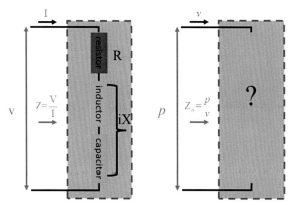

Figure 2.3 (Left) Electrical black box. (Right) Acoustical black box.

impedance Z consisting of a resistor, inductor, and capacitor are shown. The dashed line indicates the overall impedance as a "black box," an old engineering term from the early days of radio. This term implies that a more general representation of components could be inside the black box instead of just the simple arrangement shown. In general, the black box impedance is complex with a real and imaginary part, represented as a function of frequency f. $Z(f) = R(f) + iX(f)$.

For acoustics, the corresponding variables are shown in the right side of Fig. 2.3. Here pressure, p_a, is applied across an acoustic black box impedance $Z_a(f)$. The variable corresponding to electrical current is particle velocity, v_a. Similar to the electrical case, acoustic impedance as a general circuit element is defined as $Z_a(f) = p_a(f)/v_a(f)$. For now, the content of the acoustic black box impedance will remain mysterious to be revealed in detail in Chapter 3. Other correspondences include analogies to electrical charge Q and the current charge derivative with respect to time, $I = dQ/dt$, which for acoustics are displacement u and particle velocity $v = \partial u/\partial t$.

In acoustics, c represents the speed of sound in a material in meters/second (m/s). Each material has a particular density, ρ (kg/m^3) and an acoustic material impedance $Z = \rho c$ in units of Rayls. Other useful parameters are wave number, $k = \omega/c$ where ω is angular frequency $2\pi f$; and intensity, I (W/m^2) for a real Z is

$$I = pv^* = pp^*/Z = vv^*Z, \tag{2.5}$$

where the * symbol means complex conjugate, which is a mathematical operation such that if, for example, $p = p_{real} + ip_{imag}$, then the conjugate reverses the sign of the imaginary part, $p^* = p_{real} - ip_{imag}$. The corresponding electrical intensities for real Z are

$$I = VI^* = VV^*/Z = II^*Z. \tag{2.6}$$

These analogs are summarized in Table 2.1 Extensive tabulations of acoustic and elastic properties for medical ultrasound have been tabulated on the Onda website

Table 2.1 Electrical and acoustical analogs.

Sound Liquid			Sound Solid			Electrical		
Variable	Symbol	Units	Variable	Symbol	Units	Variable	Symbol	Units
Pressure	P	MPa	Stress	T	MPa	Voltage	V	Volts
Particle velocity	v	m/s	Particle velocity	v	m/s	Current	I	Amps
Particle displacement	u	m	Particle displacement	u	m	Charge	Q	Coulombs
Density	ρ	kg/m^3	Density	ρ	kg/m^3			
Longitudinal speed of sound	c_L	m/s	Longitudinal speed of sound	c_L	m/s	Wave speed	c	m/s
Longitudinal impedance	$Z_L(\rho c_L)$	Mega Rayls	Longitudinal impedance	$Z_L(\rho c_L)$	Mega Rayls	Impedance	Z	Ohms
Longitudinal wave number	k_L	m^{-1}	Longitudinal wave number	k_L	m^{-1}			
			Shear vertical speed of sound	c_S	m/s			
			Shear vertical impedance	$Z_S(\rho c_S)$	Mega Rayls			
			Shear vertical wave number	k_S	m^{-1}			

(n.d.; OndaCorp.com) and in Duck (1990) and in Appendix D. Material values for nondestructive testing such as metals and ceramics and plastics are also available, as are parameters for geophysics.

2.3 Types of waves

2.3.1 Types of propagating wavefronts and the Expanding Waves Simulator

There are three idealized wave geometries in common use. Each of three choices at the top of the Expanding Waves Simulator graphical user interface (GUI) activates an animation of an expanding or propagating wavefront. The first, the plane wave, is utilized most often, is shown in the left side of Fig. 2.4. In fact, we have already discussed this case; the plane wave in the left of Fig. 2.4, is described as $\exp\left[i2\pi f_0\left(t - x/c_0\right)\right]$, where c_0 is the medium sound speed. It is visualized, through the aid of the Expanding Waves Simulator, as an infinitely wide plane along y and z moving along the x axis. A wavefront usually has a short time extent, so for this simulator, the other extreme wave type, the impulse, $\delta\left[2\pi f_0\left(t - x/c_0\right)\right]$ is more convenient. This plane wave, when visualized from above at the position $z = 0$, is shown in the companion diagram. Note the similarity of this $z = 0$ figure in the left side of Fig. 2.2, since $t = x/c_0$, there is an equivalence between the time and space dimensions, and both the plane wave and impulse are moving toward the right along the propagation axis. In this first configuration, a plane infinite in extent moves along the one-dimensional x axis. This is an idealization of a wave from a finite sized and more realistic plane source. For a better approximation of a real device, a plane wave is multiplied by a window or cutout corresponding to the shape of the transmitter in the source plane at time zero. For example, imagine a series of plane waves propagating in time and having a rectangular or circular shape along the y and z axes.

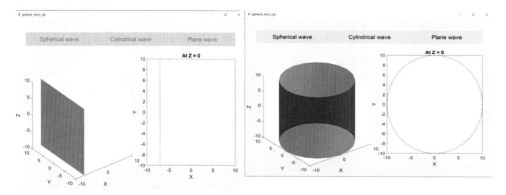

Figure 2.4 Expanding Waves Simulator: (left) plane wave, (right) cylindrical.

The two other wavefront geometries in this simulator are cylindrical and spherical. In the cylindrical case, the transmitter is an infinitely long line source along the z axis. In each plane perpendicular to the z axis is a circularly shaped expanding wavefront which has a radius, $r_{xy} = \sqrt{x^2 + y^2}$. The expanding wavefront is best described by the function, $\delta\left[2\pi f_0\left(t - r_{xy}/c_0\right)\right]$. Here, a one-dimensional source creates a wavefront changing in three dimensions, as illustrated in the right side of Fig. 2.4. In practice, this infinitely long cylinder is windowed to a finite length source of length L, resulting in a finite length cylinder. The other extreme description of this type of wavefront for a continuous wave is $\exp\left[i2\pi f_0\left(t - r_{xy}/c_0\right)\right]$.

Finally, for the spherical wavefront the emitter is a so-called point source, or a tiny three-dimensional spherical source which produces a three-dimensional wavefront expanding with a radius r in all directions. The appropriate equation for the animation is $\delta\left[2\pi f_0\left(t - r/c_0\right)\right]$ where $r = \sqrt{x^2 + y^2 + z^2}$. The continuous wave spherical wavefront expression is $\exp\left[i2\pi f_0\left(t - r/c_0\right)\right]$.

2.3.2 k-Rays

Because the sound speed is the same in any direction in media such as air or fluid, the media are called "isotropic." In general, waves in elastic media are not always isotropic. In solid materials, different types of vibrations occur and it is necessary to make distinctions to describe their behavior. In particular, the direction of propagation and the vibration (displacement) of a wave may not be in the same direction. The magnitude of a $\mathbf{k}$ vector is

$$|\mathbf{k}| = k = \omega/c = 2\pi f/c. \tag{2.7}$$

A vector is similar in concept to a complex number in which the real and imaginary parts are assigned two axes perpendicular to each other, as shown in Fig. 2.1. A vector is a convenient way of showing the direction of a "ray." The geometry for a two-dimensional $\mathbf{k}$ vector is shown in Fig. 2.5. Vectors often are shown in a **bold** format

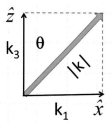

Figure 2.5 k-Ray with its magnitude $|\mathbf{k}|$ along a direction angle and projections along the $\hat{x}$ and $\hat{z}$ axes.

and $\hat{\mathbf{x}}$ is a unit vector (its magnitude $= 1$) lying the x axis and $\hat{\mathbf{z}}$, along the z axis. The expressions for the $\mathbf{k}$ vector may start to look familiar:

$$\mathbf{k} = \hat{\mathbf{x}}k_1 + \hat{\mathbf{z}}k_3 = \hat{\mathbf{x}}|\mathbf{k}|\sin\theta + \hat{\mathbf{z}}|\mathbf{k}|\cos\theta, \tag{2.8}$$

$$|\mathbf{k}| = \sqrt{k_1^2 + k_3^2}, \tag{2.9}$$

$$\theta = \arctan\left(\frac{k_1}{k_3}\right). \tag{2.10}$$

For three dimensions, the same concept is extended by including the $\hat{\mathbf{y}}$ axis, using directional cosines, as shown in Fig. 2.6:

$$\mathbf{k} = \hat{\mathbf{x}}k_1 + \hat{\mathbf{y}}k_2 + \hat{\mathbf{z}}k_3 = \hat{\mathbf{x}}|\mathbf{k}|\cos\theta_1 + \hat{\mathbf{y}}|\mathbf{k}|\cos\theta_2 + \hat{\mathbf{z}}|\mathbf{k}|\cos\theta_3, \tag{2.11}$$

$$|\mathbf{k}| = \sqrt{k_1^2 + k_2^2 + k_3^2}, \tag{2.12}$$

$$\theta_x = \arctan\left(\frac{k_1}{|\mathbf{k}|}\right), \quad \theta_y = \arctan\left(\frac{k_2}{|\mathbf{k}|}\right), \quad \theta_z = \arctan\left(\frac{k_3}{|\mathbf{k}|}\right), \tag{2.13}$$

where the angles are from each of the three cartesian axes to the three-dimensional vector $\mathbf{k}$, and k_1, k_2, and k_3 are the projections of the $\mathbf{k}$ vector on these axes.

2.3.3 Elastic waves and the Elastic Wave Simulator

Waves in gases and fluids propagate as compressional waves (the wave compresses and expands the medium as it passes by) with a single velocity, in a longitudinal mode (particle motion is in the same direction as the wave propagates). The speed of sound in water, for example, can be calculated as follows:

$$c = \sqrt{\frac{\gamma B_T}{\rho}}, \tag{2.14}$$

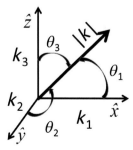

Figure 2.6 Three-dimensional directional cosines.

where $\gamma = 1.004$ is the ratio of specific heats for water, B_T is the bulk modulus, which for water at $20°C$ is $2.18 \times 10^9 \, N/m^2$, and the density $p = 998 \, kg/m^3$ so that Eq. (2.14) gives a value of 1481 m/s.

The Elastic Wave Simulator depicts a longitudinal wave to show how it differs from other types of waves. It shows the motion of different waves as animations which are reproduced here only as snapshots in Fig. 2.7. Waves are usually assumed to be longitudinal in ultrasound. Elastic waves have sound speeds related to elastic or stiffness constants and for longitudinal waves:

$$c_L = \sqrt{\frac{C_{11}}{\rho}}, \tag{2.15}$$

in which C_{11} is the longitudinal elastic constant. As shown in the bottom of Fig. 2.7, the sinusoidal motion occurs along the axis of propagation as described by $\sin(a)$ in Section 2.1.2. The material is compressed or rarified (thinned out) as the wave

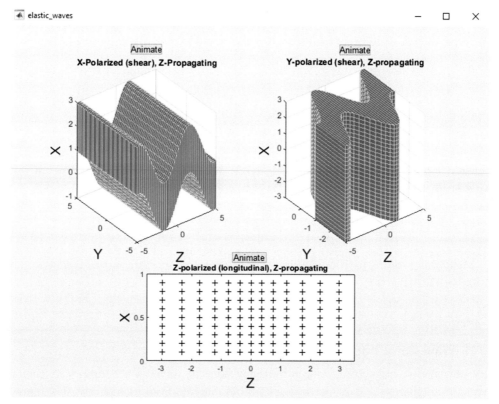

Figure 2.7 Elastic Waves Simulator animations: (upper left) shear vertical wave, (upper right) shear horizontal wave, and (bottom) longitudinal wave.

propagates. Equations for the displacement vector (particle motion) for this wave along z are

$$\mathbf{u} = \hat{\mathbf{z}} \ \sin\left[(2\pi f_0)\left(t - z/c_L\right)\right], \tag{2.16a}$$

$$\mathbf{u} = \hat{\mathbf{z}} \ \sin(\omega_0 t - k_L z), \tag{2.16b}$$

where now $k_L = \omega_0/c_L$.

Shear waves, in contrast, vibrate with motion perpendicular to the axis of propagation. The speed of a shear wave differs from that of the compressional wave, and is expressed as

$$c_S = \sqrt{\frac{C_{44}}{\rho}}, \tag{2.17}$$

where C_{44} is the appropriate shear wave elastic constant. The shear stiffness is often smaller than the compressional stiffness because shear waves do not require volumetric oscillation; it is usually easier to bend a material than to compress it. If the direction of propagation is along z, a shear vertical wave has a sinusoidal vibration in a vertical transverse direction along x (x-polarized) as illustrated by the simulator in upper left of Fig. 2.7. The corresponding equation is

$$\mathbf{u} = \hat{\mathbf{x}} \ \sin(\omega_0 t - k_S z), \tag{2.18}$$

where now $k_S = \omega_0/c_S$. This type of motion is similar to the motion of a ripple along a rope.

Another type of shear wave motion is the shear horizontal wave (y-polarized) which is more like the movement of a snake swimming in water. A picture of this wave is in the upper right of Fig. 2.7. The corresponding equation shows that the vibration is along y rather than x as before:

$$\mathbf{u} = \hat{\mathbf{y}} \ \sin(\omega_0 t - k_S z). \tag{2.19}$$

Much more information on all kinds of waves including elastic waves can be found in Szabo, (2014a).

2.4 Oblique waves at a boundary

2.4.1 Waves at a boundary

Key principles in diagnostic ultrasound imaging concern waves at a boundary. A central theme in imaging is the existence of pulse echoes reflected from boundaries. Also important is that most of the forward wave penetrates each boundary to hit deeper obstacles; a boundary encountered between dissimilar media is shown in Fig. 2.8A. General parameters shown for the two media, labeled one and two, are wave number, impedance, and speed of sound. If the incident pressure continuous wave, p_0, is perpendicular to the boundary, then the reflected wave is RF p_0 where the reflection factor, RF, is

$$RF = \frac{Z_2 - Z_1}{Z_1 + Z_2}. \tag{2.20}$$

The part of the pressure that is transmitted is TF p_0, where TF, the transmission factor, is found from

$$TF = \frac{2Z_2}{Z_1 + Z_2}. \tag{2.21}$$

Fig. 2.8A shows reflected and transmitted waves at the interface between two media; for example, with $Z_1 = 1.327$ MegaRayls (10^6 Rayls) for fat, and $Z_2 = 1.645$ MegaRayls for muscle as the two media, RF = 0.107. Note that TF = 1 + RF. If the wave had gone from right to left, what would RF and TF be? Note in this case the reflection factor is negative which indicates that the pressure wave is inverted relative to the incident wave.

2.4.2 Refraction at a boundary

In general, waves hit a boundary at an angle as illustrated in Fig. 2.8B. Here k-rays are useful. In medium 1, vectors represent incident and reflected k-rays. The tangential k-ray components are equal along $\hat{x}$,

$$k_x = k_{1L} \sin \theta_i = k_{1L} \sin \theta_r, \tag{2.22}$$

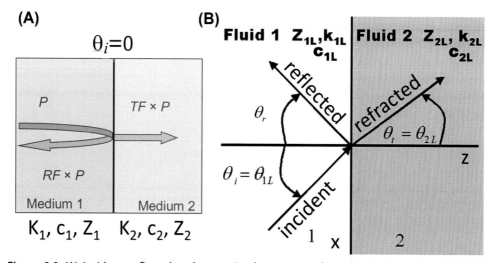

Figure 2.8 (A) Incident, reflected and transmitted waves at a boundary between two media, for normal incidence. (B) k-Rays at a boundary: incident ray at angle θ_i and reflected ray at θ_r in medium 1 and refracted ray at angle θ_t in medium 2. *From Szabo, T.L. (2014a). Chapter 3. Acoustic wave propagation. In Diagnostic ultrasound imaging: Inside out (2nd ed., pp. 67-79). Elsevier.*

where k_{1L} means the **k** vector projected onto the x-axis in medium 1, so that $\theta_r = \theta_i$. Also,

$$k_x = k_{1L} \sin\theta_i = k_{2L} \sin\theta_t, \tag{2.23}$$

where k_{2L} means **k** vector in medium 2 at angle θ_t which results in Snell's law:

$$\frac{k_x}{\omega} = \frac{\sin\theta_i}{c_1} = \frac{\sin\theta_t}{c_2}. \tag{2.24}$$

Again, $\theta_r = \theta_i$.

2.4.3 Oblique reflection and transmission at a boundary

An important effect of an oblique incident ray is that equivalent speed along z changes with angle. This change can be deduced from matching the k-rays along $\hat{\mathbf{z}}$,

$$\frac{k_z}{\omega} = \frac{\cos\theta_i}{c_1} = \frac{\cos\theta_t}{c_2}. \tag{2.25}$$

The effective or projected speed along z of sound is $c_{\theta_i} = c_1/\cos\theta_i$. Then the effective impedances as a function of angle become

$$Z_{1\theta} = \frac{\rho_1 c_1}{\cos\theta_i}, \tag{2.26a}$$

$$Z_{2\theta} = \frac{\rho_2 c_2}{\cos\theta_t}. \tag{2.26b}$$

When these impedances are substituted in Eqs. (2.20) and (2.21), the RF and TF equations for the oblique rays become

$$\mathrm{RF}(\theta) = \frac{Z_{2\theta} - Z_{1\theta}}{Z_{2\theta} + Z_{1\theta}} = \frac{Z_2 \cos\theta_i - Z_1 \cos\theta_t}{Z_2 \cos\theta_i + Z_1 \cos\theta_t}, \tag{2.27a}$$

$$\mathrm{TF}(\theta) = \frac{2Z_{2\theta}}{Z_{1\theta} + Z_{2\theta}} = \frac{2Z_2 \cos\theta_i}{Z_2 \cos\theta_i + Z_1 \cos\theta_t}. \tag{2.27b}$$

Similarly, equations for power transfers across the boundaries can be derived, first, the reflected power ratio:

$$rfp = \left| \mathrm{RF}(\theta) \right|^2, \tag{2.28a}$$

and then the transmitted power ratio,

$$tfp = \frac{4Z_1 Z_2 \ \cos\theta_i \ \cos\theta_t}{(Z_2 \ \cos\theta_i + Z_1 \ \cos\theta_t)^2},$$

(2.28b)

where it is convenient to know that power is conserved:

$$rfp + tfp = 1.$$

(2.29)

2.4.4 Oblique simulator

The Oblique Refraction Simulator, found in Lecture 2 of the Control Board, is based on the analysis of the previous section. Any combination of materials can be selected for medium 1 or 2 from dropdown menus. The k-rays are color coded according to the colorbar on the right which represents values for zero to one as illustrated in Fig. 2.9. Note that the lengths of the **k** vectors are not shown to scale, for clarity. The colors encode the ratio of power either being reflected or transmitted; see Eqs. (2.28a) and (2.28b). The angle of incidence is adjustable and has a default value of 0 degrees. The angle of refraction is determined from Eq. (2.24). One can easily observe that less and less power is transmitted as the incident angle increases. Finally, when the angle of

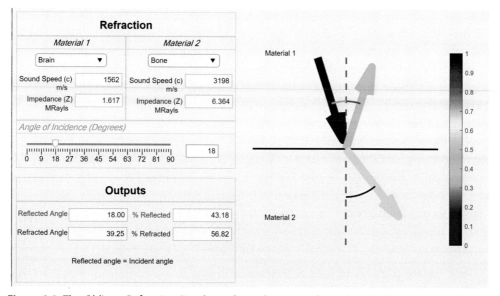

Figure 2.9 The Oblique Refraction Simulator shows k-rays at a boundary: incident ray at angle controlled by slider control. The k-rays are color-coded by colorbar scale at right with numerical power values given by output boxes. Media 1 and 2 are selected by value pulldown menus and numbers are filled in after selection.

transmission θ_t becomes 90 degrees, the critical angle is reached and all of the power is reflected. Note that at this angle $\theta_{i\text{-}critical}$ and beyond, the value of tfp in Eq. (2.28b) and the k-ray along $\hat{\mathbf{z}}$ for medium 2 are zero.

In a more typical example of a tissue boundary than that shown in Fig. 2.9, fat (medium 1) and muscle (medium 2), only 1% of the power is reflected, and 99% is transmitted as given quantitatively from Eq. (2.28) in the output section. This favorable transmission ratio is one of the reasons why diagnostic ultrasound succeeds: a small reflection at the boundary is detected as a reflection or pulse echo and most of the rest of the power is free to intercept deeper boundaries.

Much more information on this topic can be found in Szabo, (2014a). Although throughout most of this book we assume waves are of the longitudinal type, other types exist as we know from Section 2.3.3. In general, both longitudinal waves and shear waves (and others!) can be created at a boundary but because of the complexities involved, these combinations are not covered here and the reader is referred to Szabo (2014a) for more details and computational models.

2.5 Pulses reverberating in layers

2.5.1 Reverberations in a layer

What happens to acoustic waves when there is more than one boundary? In other words, waves in a layer or layers? To investigate this configuration, we bring back the other extreme type of wave, the short pulse, and invoke the relationships established in Section 2.4.1. If we revisit those equations and subscripts 1 with $m = 1$ and 2 with $n = m + 1$, we have general descriptions for acoustic waves at a boundary and can use them to index more than one boundary. For example, we can rename Eqs. (2.20) and (2.21) as RF_{mn} and TF_{mn} where in the previous context, $m = 1$ and $n = 2$. With reference to the top of Fig. 2.10, we have three materials, numbered 1, 2, and 3 and two boundaries between them. A wave from material 1 at point A passes into material 2, then through the second boundary and into material 3 at location B. If the layer has a width w, then the impulse with amplitude p_0 passes the first boundary and becomes $\mathrm{TF}_{12}\,p_0\,\delta(t)$. At point B, there is a delay from crossing the width w corresponding to $t = w/c$, and the pulse emerges from the second boundary as

$$p_1 = \mathrm{TF}_{12}\cdot\mathrm{TF}_{23}\cdot p_0\delta\left(t - w/c\right). \tag{2.30}$$

Now we are ready to include reflections as shown in the bottom of Fig. 2.10. An impulse starting at point A goes to point B where it separates through transmission into material 3 as before and a reflection which heads for point C. At C, part of the impulse passes through the first boundary and the rest is reflected toward point D. At point D, there is a transmitted and reflected part, and so on. If only the transmitted

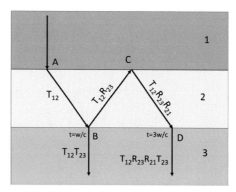

Figure 2.10 Graphical illustration of a vertically oriented reverberation for a layer (2) bounded by semiinfinite media 1 and 3. A short pulse passes through the boundary from 1 and emerges in 2 as T_{12} and passes into 3 as $T_{12}T_{23}$ at B where it is reflected and travels to 3 at D. Pulses transmitted into 3 occur at time delays of $t = w/c$ and $t = 3w/c$ with c the speed of sound in 2. The angles indicated by the "rays" are not physically accurate; rather they are used to illustrate the wave interactions with the interfaces by using exaggerated angles.

waves through the second boundary are of interest, the overall result is a combination of all the factors along the path. There are transmissions at A and D and reflections at B and C, so the total description for the impulse emerging at D is

$$p_3 = \mathrm{TF}_{12} \cdot \mathrm{RF}_{23} \cdot \mathrm{RF}_{21} \cdot \mathrm{TF}_{23} \cdot p_0 \delta\big(t - 3w/c\big). \tag{2.31}$$

Note for graphical clarity, the path shown in Fig. 2.10 is zigzag (slanted paths), but it is really all vertical (up or down), perpendicular to the boundaries. If it was a tilted path, then we would have to use Eq. (2.27) generalized to m and n and a path length of $w/\cos\theta_{t2}$ with θ_{t2} being the transmitted angle in material 2, so for simplification, we assume a perpendicular path. If the transmissions through boundary 2 are added (i.e., p_1 and p_3 at times $t = w/c$ and $t = 3w/c$), the result is a time series of transmitted pulses. This process can be continued for more internal reflections.

2.5.2 Layer Pulse Simulator

The previous section has laid the groundwork for the Layer Pulse Simulator, or Pulsed Pressure Wave Reverberation Simulator in Lecture 2 of the Control Board, which is shown in Fig. 2.11. The familiar graph represents the reverberation perpendicular paths as a zigzag pattern. The impulse starts at the top and propagates downward with an initial amplitude of 1. The top material 1 and the bottom material 3 are assumed to be semi-infinite and their acoustic impedances are selected in the upper left of the GUI. For the layer, material 2, its impedance, speed of sound (mm/μs), and thickness in mm are all selectable. The output, shown at the lower right of the GUI is a plot of

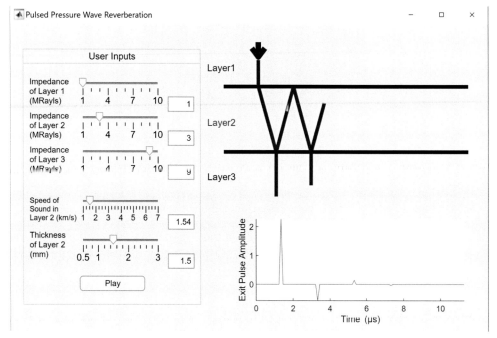

Figure 2.11 The Layer Pulse Simulator indicates the transmission of pulses through a middle layer. Impedances of three layers are selectable as are the thickness and speed of sound of layer 2. The pulses that are geometrically propagating up and down, are shown in a zigzag pattern so the pulses' progress can be indicated in an animation.

the time signature of all the pulses transmitted through boundary 2. Once the calculations are initiated, a bright spot travels along the zigzag path as animation while the time signals are traced out below. The many possible combinations of parameters provide opportunities to achieve different objectives: to maximize throughput for different constraints or to shorten the number of reverberations. More information on these effects can be found in Szabo, (2014b) and Szabo (2010). Later, this approach will play an important role in the understanding of matching layers and transducers.

2.6 Waves in layers

2.6.1 Continuous waves in a layer

At the other extreme, the behavior of a continuous wave in a layered medium is very different than that of a short pulse. To understand this difference, refer to Fig. 2.12, where we seek an expression for Z_{in1}. Consider a pressure wave p_0 from medium 1 traveling past the first boundary of the layer with a width w and an impedance, Z_2, into a third impedance, Z_3. A plane wave going to the right in material 2 is $\exp(\omega t - k_2 z)$ and a plane

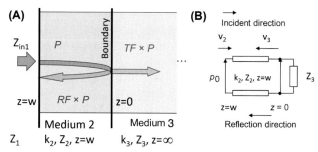

Figure 2.12 Input impedance through a layer. (A) A continuous pressure wave p_0 from semi-infinite medium 1 enters medium 2, a layer of thickness w which is loaded by a semi-infinite medium 3 of impedance Z_3. A reflection occurs at the boundary at $z = 0$ as well as a transmission. (B) A model diagram at the right shows particle velocities and reflection in the layer. *Adapted from Szabo, T. L. (2014a). Chapter 3. Acoustic wave propagation. In Diagnostic ultrasound imaging: Inside out (2nd ed., pp. 67-79). Elsevier.*

wave going right to left is $\exp(\omega t + k_2 z)$, so summing the forward wave and its reflection from the second boundary and material 3:

$$p(\omega) = p_0 e^{i\omega t}\left[e^{-ik_2 z} + \mathrm{RF}_{23}e^{ik_2 z}\right], \tag{2.32}$$

and

$$v(\omega) = \frac{p_0}{Z_2}e^{i\omega t}\left[e^{-ik_2 z} - \mathrm{RF}_{23}e^{ik_2 z}\right] = v_2(\omega) - v_3(\omega), \tag{2.33}$$

where

$$\mathrm{RF}_{23} = \frac{Z_3 - Z_2}{Z_2 + Z_3}, \tag{2.34}$$

results in an input impedance as seen from medium 1 as

$$Z_{in1}(\omega) = \frac{p(\omega)}{v(\omega)}, \tag{2.35}$$

which, after combining Eqs. (2.33), (2.34), and (2.35), becomes

$$Z_{in1}(\omega) = Z_2\left(\frac{Z_3 \, \cos k_2 w + iZ_2 \, \sin k_2 w}{Z_2 \, \cos k_2 w + iZ_3 \, \sin k_2 w}\right). \tag{2.36}$$

Instead of the familiar constant impedance, this equation implies that the input impedance varies with frequency through $k_2 = 2\pi f/c_2$ and the width w of the layer. Alternatively, this impedance depends on $k_2 w = 2\pi(w/\lambda_2)$, the layer width in wavelengths. This result will lead to strange effects as explored with the simulator.

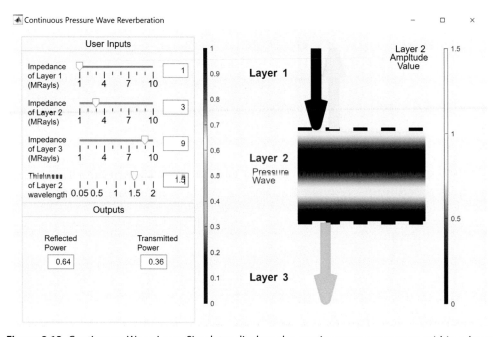

Figure 2.13 Continuous Wave Layer Simulator displays the continuous wave pattern within a layer surrounded above and below by semi infinite media whose impedances are all selectable. The thickness of the layer in wavelengths can be varied as an input. k-Rays normal to the layer show the power reflected and transmitted through layer 2 according to the colorbar to the left and as numerically shown in the output section. The pressure amplitude distribution within the layer is given by the colorbar to the right.

2.6.2 Continuous Wave Layer Simulator

The setup on the Continuous Wave Layer Simulator, in Lecture 2 of the Control Board, has the inputs in the upper left quadrant of the GUI. These inputs include the impedances of the three materials, the middle one having a width in wavelengths and the other two being semiinfinite in extent (see Fig. 2.13). In the color graphic, vectors for incident, reflected, and transmitted power are shown for normal incidence, and they use the same color coding for the same fraction of power scale (zero to one) used on the Oblique Refraction Simulator. Inside the layer, the continuous wave pattern is color coded according to the scale on the right.

In this simulator the impedance of material 1 comes into play as well as the variables associated with the layer. Material 3 can be thought of as a load. Overall, the input impedance is really a function of w/λ_2, as well as Z_2 and Z_3. The default configuration has the impedances 1, 3, and 9 Mega Rayls and a width of $3\lambda/2$. This configuration has the effect of making the layer disappear. For this case, Eq. (2.36) yields a value of $Z_{in} = Z_3$; therefore, this situation is the equivalent to having a power transfer for material 1 directly adjacent to material 3 according to Eq. (2.28).

For maximum power transfer, $w = \lambda/4$. Here,

$$Z_{in1} = Z_2^2/Z_3. \tag{2.37}$$

This width in wavelengths and the choice of selecting a value of Z_2 for the layer impedance according to the formula, $Z_2 = \sqrt{Z_1 Z_3}$ creates a "matching layer" for maximum power transfer. This arrangement results in $Z_{in1} = Z_1$ instead of the complicated Eq. (2.36), for a perfect transfer according to Eq. (2.21). Many other interesting properties can be explored using this simulator. More information can be found in Szabo, (2014b).

2.7 Lab 2: reflection and refraction of acoustic waves

2.7.1 Physics simulator applications

The simulators are introduced in Section 1.4. This collection of GUI-based programs is available for an additional fee and can be installed on a personal computer using either Windows 10 or Mac OS X. Please see the website Verasonics.com for further information on computing requirements and compatibility. Though the simulators were developed using MATLAB® scientific software, the simulator applications are in an executable form which does not require a MATLAB® license. More details are available at Verasonics.com.

Detailed information about each simulator program is provided in the simulator manual that comes with the simulator software. For each program, the manual presents a list of the input and output variables and describes output display options if appropriate.

2.7.2 Lab 2 learning objectives

The exercises with the physics simulators are intended to introduce the student to the nature of wave propagation in uniform and in layered media. First, the student is encouraged to simply explore what happens by "playing" with the controls on the simulator GUI panels described earlier in this chapter. The student gains an intuitive appreciation for different wave geometries by observing plane, cylindrical, and spherical wave propagation in uniform media, visualized by animating the motion of a wavefront with time. Propagation in a layered medium is explored using both continuous and pulsed waves (see Sections 2.5.2 and 2.6), first presenting interaction at a single interface through simulation of refraction following Snell's law and then observing the consequence of multiple reflections between boundaries. Reflection and transmission coefficients as a function of medium parameters and incident angle are introduced and visualized using the graphical tool, and the student is guided to discover special cases such as total internal reflection by simply adjusting parameters on a control panel rather than by manipulating equations. The relevant equations using medium impedance are provided in this chapter for use by enterprising students who wish to associate the physics with analytical forms.

Concepts of multiple internal reflections and reverberation signals are explored using the continuous wave and pulsed wave simulators and these tools will be used in explaining observations made using the imaging system and small custom acoustic phantoms in the practical laboratory exercises in this chapter as well as in Chapter 4 where transducers are seen as a special type of layered medium. The counter-intuitive properties of quarter-wave and half-wave layers are explained and demonstrated using the simulators.

Several concepts are revisited using the research imaging system and small custom acoustic phantoms, laboratory kits (see Fig. 1.23), including observation of refraction effects on the apparent position of a target viewed through a thick layer using a block of acrylic (plexiglass) within a water bath. This introduction to aberrating media makes clear that such apparent shifts in depth or lateral position can be very significant and is an important lesson for sonographers and ultrasound engineers alike. Later in Chapter 3, reverberant echo signals are generated using a variety of thin layers. The concepts are further explored with different waveforms and observations of multiple reflections.

2.7.3 Lab 2 description of exercises and illustrative results

The first experiment using the Verasonics (n.d.) Vantage™ research ultrasound system uses the "refractor" insert in the water tank, illustrated in Fig. 2.14. The goal is to determine the sound speed of a layer by a replacement approach, that is, observing the difference in travel time to a deeper target when propagating through the layer and when propagating through the surrounding (uniform) medium alone. The phantom is made using a solid layer with parallel interfaces that has a faster compressional sound speed than the water. Multiple reflections occur from the top and bottom surfaces as described in Section 2.5.2. The layer has a given thickness but has an unknown sound speed. Pin targets below the layer and oriented parallel to the layer interfaces are visible to the transducer either through a purely waterborne path, or through the layer; therefore, for this experiment, as the pins as seen through a water path, and then through the refractor, the pins appear to move vertically because of the changes in sound speed along the acoustic path.

Once the transducer is properly aligned with the phantom, one can see a number of horizontal line reflections that are important for the student to identify. The student is also able to quantify the apparent vertical shift of the pins between the pure water path and the layered medium path, as observed in Fig. 2.15, then uses this shift to estimate the sound speed in the layer of given thickness. Some rudimentary travel time equations are provided in the laboratory lecture notes to assist in the process.

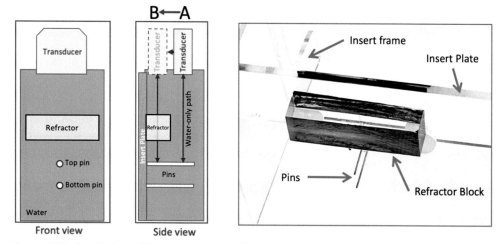

Figure 2.14 The "refractor" insert is diagramed on the left in face and side views, and is depicted in a photograph on the right. The refractive layer is acrylic with plane parallel polished interfaces on top and bottom. The pins are stainless steel with a diameter of just under a millimeter. They present bright, nearly point-like targets to the imaging system when the transducer is parallel to the insert mounting plate.

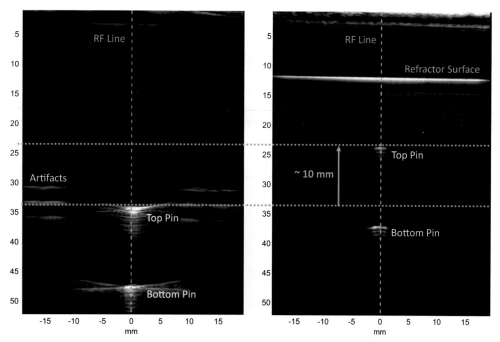

Figure 2.15 Images showing vertical shift of pin echoes with identifying notations, first in water (left) and with the refractor material inserted in the path (right).

A single RF line can be plotted to get precise estimates of the echo time shift, as illustrated in Fig. 2.16. The plot includes the reconstructed RF line data at defined pixel locations, which is relatively sparse, the interpolated curve to show the detailed oscillatory nature of the underlying RF field, and the envelope plot to facilitate travel time estimation.

The second experiment uses the same physical setup but looks at apparent lateral shift resulting from oblique angles of incidence. A set of steering angles can be explored by using a slider on the GUI control panel to change the steering angle of the acoustic beams (beams and beamforming are discussed in Chapters 5 and 6). By changing the angle and measuring the lateral shift, the student can compare observations with simulations of refraction using Snell's Law, or their own computations. An example of lateral shift appears in Fig. 2.17 for two oblique angles.

A complication arises from the fact that the image of the pin targets degrades significantly when refraction is present. This is a result of beamforming using an inaccurate model for the medium sound speed, an issue that will be explored in detail in Chapter 8; the pins viewed through the refractor are now "out of focus" because the arrival time of the echoes does not align with the geometry expected using a water sound speed. This effect is so severe, given the large difference between the sound speeds in water and acrylic, that the arc representing the poorly focused echo using the sound speed in water is very difficult to use to measure lateral shift. For this

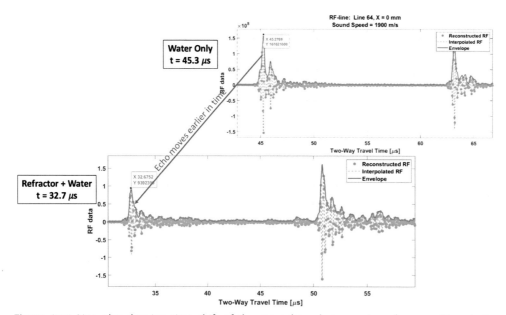

Figure 2.16 Line plot showing time shift of the pin echoes between transducer position A, the water-only path (top) and position B, the refractor plus water path (bottom).

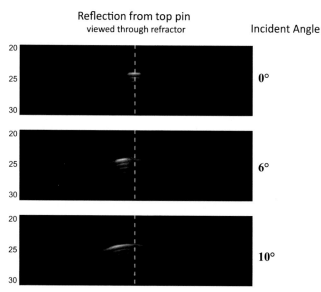

Figure 2.17 Images showing lateral shift of pin echoes when steering the beam and using the thick refractor. The central green line normally used to collect RF data is here simply used as a reference.

experiment, the discerning student may notice that we use a compromise value of sound speed that partially compensates for the mismatch and thus simplifies determining the shift when using the "Refractor" setting on the GUI control panel.

References

Duck, F. A. (1990). *Physical properties of tissue: A comprehensive reference book*. Academic Press.

Szabo, T. L. (2014a). Chapter 3. Acoustic wave propagation. In Diagnostic ultrasound imaging: Inside out (2nd ed., pp. 67−79). Elsevier.

Szabo, T. L. (2014b). Transducers. In Diagnostic ultrasound imaging: Inside out (2nd ed., pp. 142−148). Elsevier.

Onda website. (n.d.). *Acoustic tables*. https://www.ondacorp.com/resources/. Accessed July 25, 2022.

Szabo, T. L. (2010). *Medical ultrasound sensors*. In D. Jones (Ed.), *Biomedical sensors*. Momentum Press.

Verasonics. (n.d.). https://verasonics.com/. Accessed July 25, 2022.

CHAPTER 3

Signals

3.1 Overview

3.1.1 Play with blocks

Fig. 3.1 shows a basic functional diagram for an imaging system. Electrical signals are shown in gray. An electrical waveform is created in the first block E. This signal eventually enters the transducer shown in green which, acting as a transmitter block, G_T, converts the waveform into an acoustic pressure signal. The pressure waveform travels into an absorbing medium, represented by block, A_T and hits a scatterer S. The pulse

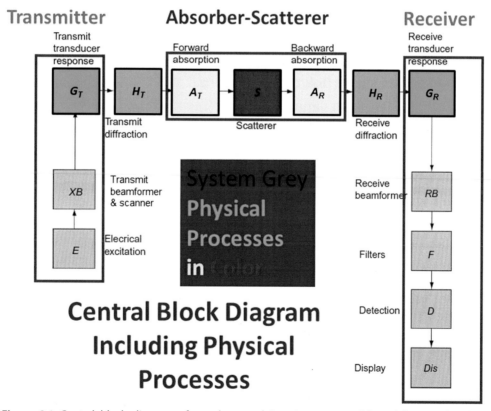

Figure 3.1 Central block diagram of an ultrasound imaging system. *Adapted from Szabo, T. L. (2014a). Overview. In Diagnostic ultrasound imaging: Inside out (2nd ed.). Elsevier.*

Essentials of Ultrasound Imaging
DOI: https://doi.org/10.1016/B978-0-323-95371-9.00006-0

echo signal is reflected and sent back, through a return path, A_R, to the transducer for reception, G_R, where it is reconverted into an electrical waveform. Finally, this waveform passes through a set of filters, F.

3.1.2 In this chapter you will learn

The ultrasound imaging system block diagram represents the system as a series of functional blocks that each transform the arriving information before sending it on to the next block. Several numerical simulators are used to explore the transformations. The characteristics of different types of electrical waveforms and their corresponding spectra can be examined using the Fourier Transform Simulator. Many blocks act as filters, and the Fourier Filter Simulator allows exploration of different waveforms and their combination with many filters. Electrical circuit elements can be represented by small matrices, and their use is demonstrated by the *ABCD* Simulator which assembles various electrical circuit components. Acoustic signals also undergo linear filtering effects and can be handled in a similar way. But in acoustics, the absorption filter varies with the propagation medium or tissue type and is distance dependent, as illustrated by the Absorption Simulator.

3.2 Fourier transforms link time waveforms and frequency spectra

3.2.1 Fourier Transform Simulator

The Fourier Transform Simulator introduces waveforms and their spectra. The oscillatory blue line in the top right plot of Fig. 3.2 is an example of a real-time waveform: a Gaussian pulse, $p(t)$. The signal is a continuous wave windowed by a Gaussian envelope. Plotted below the pulse is its spectral magnitude, $|P(f)|$. The two displayed functions are related by the mathematical operation called the (minus i) Fourier transform (Bracewell, 2014; Szabo, 2014a, 2014b) defined by

$$P(f) = \mathfrak{I}_{-i}[p(t)] = \int_{-\infty}^{\infty} p(t)e^{-i2\pi ft}dt. \tag{3.1}$$

In general, $P(f)$, is complex. The time waveform can be obtained from $P(f)$ by an inverse Fourier transform,

$$p(t) = \mathfrak{I}_{-i}^{-1}[P(f)] = \int_{-\infty}^{\infty} P(f)e^{i2\pi ft}df. \tag{3.2}$$

These forms $P(f)$ and $p(t)$ are alternate but completely equivalent ways of describing the same signal. Some problems are more easily solved in the frequency domain, while others are better addressed in the time domain.

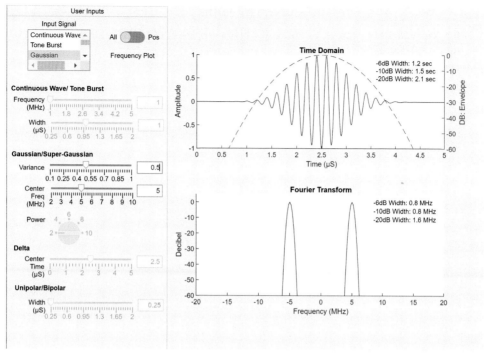

Figure 3.2 Fourier Transform Simulator. Input variable panel on the left, with output plots in the time and frequency domains on the right.

For the Gaussian pulse shown in the simulator of Fig. 3.2,

$$g(t) = g_0 \exp[-w(t-t_d)^2]\cos(2\pi f_0 t), \tag{3.3}$$

where t_d is a time delay, the waveform's corresponding spectrum is

$$G(f) = \sqrt{\pi/w}\ \exp(-i2\pi f t_d)\exp\{-(\pi^2/w)[(f-f_0)^2 + (f+f_0)^2]\}. \tag{3.4}$$

The pulse shown in the simulator is a cosine wave of center frequency 5 MHz amplitude modulated by a Gaussian function with a width $w = 0.50$. As shown in the waveform window, the pulse is centered so that it is delayed by $t_d = 2.5$ μs. The spectral magnitude plotted in the spectrum window on a logarithmic scale has two Gaussian-shaped parts, one centered at $f = f_0$ and one at $f = -f_0$. Is there such a thing as a negative frequency? Yes, in this case, it does exist mathematically as a consequence of the Fourier transform operation on the specific $g(t)$ given by Eq. (3.3). Finally, the widths of pulses and spectra in the Fourier Transform Simulator provide quantitative data for trial experiments and comparisons.

3.2.2 Hilbert transform and pulse envelope

Another signal feature that is essential for imaging is called the "analytic envelope." To obtain this envelope, a mathematical operation called the "Hilbert transform" (Szabo, 2014b) is performed on $g(t)$ which turns a real signal into a quadrature one:

$$\tilde{g}(t) = H_i \big[g(t) \big]. \tag{3.5}$$

Often, the quadrature signal is formed from its radio frequency part shifted by 90 degrees in phase from that in the real signal. For example, for the real $g(t)$ of Eq. (3.3), it is

$$\tilde{g}(t) = g_0 \exp \big[-w(t - t_d)^2 \big] \sin(2\pi f_0 t). \tag{3.6}$$

An analytic complex signal combines the real and quadrature signals:

$$a(t) = g(t) + i\tilde{g}(t), \tag{3.7}$$

so that the pulse analytic signal envelope, $e(t)$, can be constructed:

$$e(t) = \sqrt{g(t)^2 + \tilde{g}(t)^2}, \tag{3.8}$$

which for this example is

$$e(t) = g_0 \exp[-w(t - t_d)^2]. \tag{3.9}$$

Returning to Fig. 3.2, a dashed red curve represents the envelope of this waveform, plotted on a dB scale (shown on the right—note that the red envelope would follow the peaks of the blue waveform if they were plotted on the same scale). Commonly, envelope widths are given for this waveform and provide a quantitative way of comparing different signal shapes.

Similarly, bandwidths at different dB levels are given for an individual (centered at f_0) frequency representation in the spectrum panel. The fractional bandwidth in percent is defined as the difference in band edge frequencies at a −6 dB level (absolute bandwidth, normalized by f_0), all multiplied by 100. For the example shown in Fig. 3.2, the absolute −6 dB bandwidth is 0.8 MHz, and the corresponding fractional bandwidth is 16%.

The Gaussian function provides a means of producing a compact envelope in the time domain which is applicable for high resolution imaging and we will return to optimum waveforms for imaging later in this chapter. Other waveforms already discussed, such as the impulse and continuous wave, as well as many others and their spectra, can be examined with this simulator. More information about waveforms, Fourier transforms, and Hilbert transforms can be found in Chapter 2 and Appendix A of Szabo (2014a, 2014b). Note that for this simulator, spectra and envelopes are computed using digital Fourier transforms which approximate analytical Fourier transforms.

3.3 Blocks and filters

3.3.1 Combining blocks

Combining linear system blocks such as those in Fig. 3.1 is straightforward: the frequency responses of the individual blocks are multiplied:

$$G_0(f) = E_1(f)G_2(f), \tag{3.10}$$

or equivalently, the temporal functions are convolved in the time domain (a mathematical operation denoted by $*_t$):

$$g_0(t) = e_1(t) *_t g_2(t), \tag{3.11}$$

A remarkable property of the Fourier transform is that multiplication in one domain is equivalent to convolution in the other. These operations are diagramed in Fig. 3.3.

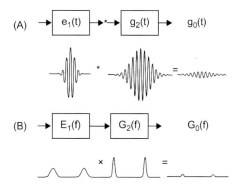

Figure 3.3 (A) Time domain block connected by convolution. (B) Frequency domain block connected by multiplication. *From Szabo, T. L. (2014b). The Fourier transform. In* Diagnostic ultrasound imaging: Inside out *(2nd ed.). Elsevier.*

3.3.2 Fourier Filter Simulator

The Fourier Filter Simulator in Fig. 3.4 combines two blocks from Fig. 3.1 with $e_1(t)$ (the excitation) functioning as an input signal and $g_2(t)$ (the transducer characteristic) as a filter. Both are selectable from dropdown menus shown in input panel on the left side of the simulator. In this example, once the main type of filter is selected, additional specific parameters can be adjusted. The final adjusted filter spectrum is shown at the bottom of the input panel. The output panel to the right displays the input pulse and its envelope, here recognized as a Gaussian. To the right, the input spectrum, $E_1(f)$, and filter spectrum, $G_2(f)$ are plotted. Finally, the resulting output time waveform, $e_0(t)$, and its spectrum (the product of the two spectra above), $E_0(f)$, are shown in the lower right output panel along with quantitative envelope and bandwidth measurements useful for comparisons.

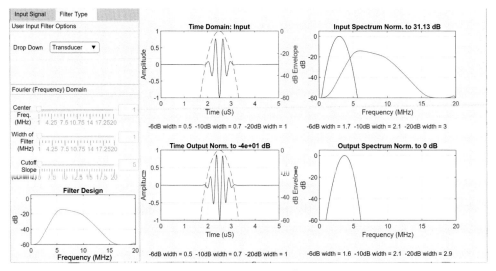

Figure 3.4 Fourier Filter Simulator panel. Input panel on left; output panel with plots and quantitative widths on right.

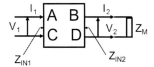

Figure 3.5 The *ABCD* network matrix and variables. The network is terminated in a load Z_M. *From Szabo, T. L. (2014c). Acoustic wave propagation. In* Diagnostic ultrasound imaging: Inside out *(2nd ed.). Elsevier.*

In this example, because an asymmetric transducer response is chosen, the output is also slightly asymmetric (though this is difficult to see in the figure), indicating a practical realization.

3.4 *ABCD* matrices

3.4.1 *ABCD* block

Another tool useful for describing linear electrical networks and acoustic waves is called the "*ABCD* matrix" (Szabo, 2014c). In Fig. 3.5 two connections in and out of a general linear network block allow the description of both variables: voltage and current. The relationships between input and output parameters in turn define a transfer function for the block and input and output electrical impedances. A matrix is a shorthand mathematical representation of simple linear equations:

$$\begin{pmatrix} V_1 \\ I_1 \end{pmatrix} = \begin{pmatrix} A & B \\ C & D \end{pmatrix} \begin{pmatrix} V_2 \\ I_2 \end{pmatrix} \tag{3.12}$$

which when multiplied out results in the equations,

$$V_1 = AV_2 + BI_2 \tag{3.13a}$$

$$I_1 = CV_2 + DI_2. \tag{3.13b}$$

The key outcomes of this approach are equations for the transfer function and input impedance:

$$\frac{V_2}{V_1} = \frac{Z_M}{AZ_M + B}, \tag{3.14}$$

$$Z_{IN1} = \frac{AZ_M + B}{CZ_M + D}. \tag{3.15}$$

Surprisingly complicated networks can be constructed from combining the four elemental blocks shown in Fig. 3.6.

Common circuit elements, individually or in combination, such as resistors, capacitors, and inductors, can be placed in either the series or shunt relationship to other circuit elements. The combination of blocks can be represented by a multiplication of *ABCD* matrices, resulting in a new overall 2×2 matrix.

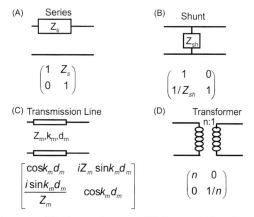

Figure 3.6 (A) Series element. (B) Shunt element. (C) Transmission line. (D) Transformer. *From Szabo, T. L. (2014c). Acoustic wave propagation. In Diagnostic ultrasound imaging: Inside out (2nd ed.). Elsevier.*

A similar two port approach can be applied to pressure, p_a, and particle velocity, v_a, variables instead of voltage and current to describe wave propagation in layers as described later, and in more detail in Szabo (2014c).

$$p_{a1} = A p_{a2} + B v_{a2} \tag{3.16a}$$

$$v_{a1} = C p_{a2} + D v_{a2}. \tag{3.16b}$$

3.4.2 Cascaded *ABCD* blocks

Let us assume that we wish to solve for the input impedance and the overall transfer function for the simple circuit in the top of Fig. 3.7. We can first represent the circuit elements as *ABCD* matrix blocks. To combine two blocks, two approaches work equally well. The first, starts at the rightmost load Z_R and finds the input impedance, Z_{IN1} using matrix 1 and Eq. (3.15) and V_2/V_3 from Eq. (3.16) with $Z_M = Z_R$. Then the process is repeated except that matrix 2 is used and $Z_M = Z_{IN1}$, and the overall voltage ratio is $V_1/V_3 = (V_1/V_2)(V_2/V_3)$.

The second approach is to multiply the matrices and then apply the two equations to the overall product matrix with $Z_M = Z_R$. Use the matrix product rule,

$$\begin{pmatrix} A_1 A_2 + B_1 C_2 & A_1 B_2 + B_1 D_2 \\ C_1 A_2 + D_1 C_2 & C_1 B_2 + D_1 D_2 \end{pmatrix} = \begin{pmatrix} A_1 & B_1 \\ C_1 & D_1 \end{pmatrix} \begin{pmatrix} A_2 & B_2 \\ C_2 & D_2 \end{pmatrix}. \tag{3.17}$$

3.4.3 *ABCD* Simulator

The *ABCD* Simulator provides the response of three *ABCD* matrices which can be populated with different circuit elements and values shown in the pink input panels of Fig. 3.8. In the upper panel, the terminating impedance values can be selected. In the

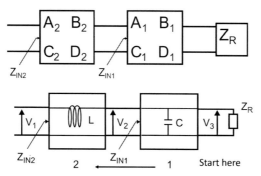

Figure 3.7 (Top) Generic cascaded blocks. (Bottom) Matrix 1: Shunt capacitor and Matrix 2: Series inductor. *From Szabo, T. L. (2014c). Acoustic wave propagation. In* Diagnostic ultrasound imaging: Inside out *(2nd ed.). Elsevier.*

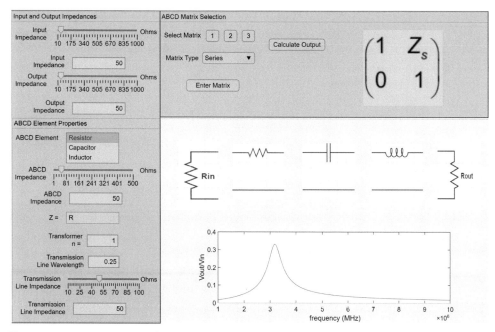

Figure 3.8 *ABCD* Matrix Simulator. (Top) Matrix type selector and matrix graphic shaded in red. (Left upper corner) Impedance input shaded in red. (Lower left) *ABCD* matrix specifics. (Lower right) Overall matrix layout and overall transfer function.

upper panel, each of the matrix types can be chosen, and an image of the type is indicated. The lower left section, specific element values can be selected. In the example shown, a series resistance of 50 Ω is chosen, and with the other elements, the circuit has a series resonance at a frequency of 3.1 MHz. In the lower right output panel, symbols of the matrix elements are connected in order and the final voltage transfer ratio is graphed as a function of frequency.

3.5 Absorption

3.5.1 Power law absorption in the frequency domain

The Absorber—Scatterer group in Fig. 3.1 consists of three blocks: a transmit path absorber, a scatterer, and a return path absorber. The scatterer can be as simple as a perfect mirror. The kind of absorption that is most common is frequency power law absorption (Holm, 2019; Szabo, 2014d) which can be described by

$$A(f, z) = A_0 \exp(-\alpha_0 |f|^\gamma z). \tag{3.18}$$

Note that this is a low pass frequency filter but it also depends on the distance z. If either the natural logarithm or the logarithm to the base 10 of this equation is taken, absorption magnitude can be expressed in units of nepers per unit distance or decibels per unit distance ($\alpha_{dB} = 8.6886\ (\alpha_{nepers})$) as depicted in Fig. 3.9. Similarly, for a fixed distance z, absorption decays exponentially as a function of frequency.

For the pulse echo or round-trip case, the magnitude of the overall Absorber—Scatterer for a mirror, RF = 1, becomes

$$\left|A_{RT}(f, z)\right| = A_0 \exp(-\alpha_0 |f|^\gamma z) RF \exp(-\alpha_0 |f|^\gamma z) = A_0 \exp(-\alpha_0 |f|^\gamma 2z). \quad (3.19)$$

For a more complete description of absorption, additional terms are needed. A time delay, encountered in Eq. (3.3), accounts for the finite propagation time due to the speed of sound, c. A phase delay term shows where in time the measurement point z is, $t = z/c$, or where a scatterer is located, $t_{RT} = 2z/c$.

Another term is needed to satisfy the conditions of causality. Causality has a very intuitive physical meaning: the arrival of energy cannot occur before the start of the event! In general, this extra term, $\beta_E(f)$, accounts for some adjustment and distortion of the pulse shape that is needed to satisfy causality when absorption occurs. Mathematically, these terms together provide the equivalent of an overall multiplication of the signal as a function of time by a step function starting at zero time, therefore assuring causality.

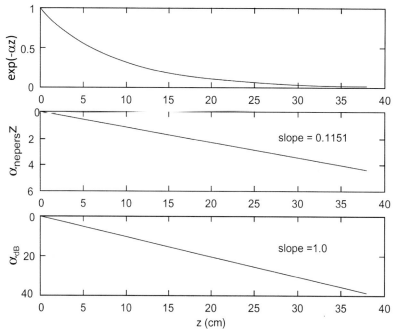

Figure 3.9 Constant absorption as a function of depth on a (top) linear scale, (middle) dB scale, and (bottom) neper scale. *From Szabo, T. L. (2014d). Attenuation. In Diagnostic ultrasound imaging: Inside out (2nd ed.). Elsevier.*

To combine all these terms together we use a material transfer function,

$$MTF(f, z) = \exp[\gamma_T(f)z],\tag{3.20}$$

where the three terms in the propagation constant γ are given as,

$$\gamma_T(f) = -\alpha(f) - i\beta(f) = -\alpha(f) - i[k_0(f) + \beta_E(f)],\tag{3.21}$$

and so, in addition to the amplitude falloff due to absorption which is a function of frequency, there is a small change to the speed of sound with frequency, $c(f)$, in addition to the expected delay due to the nominal sound speed, c_0,

$$1/c(\omega) = \beta/\omega = k_0/\omega + \beta_E/\omega = 1/c_0 + \beta_E/\omega.\tag{3.22}$$

These changes with frequency, called "dispersion," are shown for typical tissues in Fig. 3.10. $\beta_E(f)$ can be found from simple algebraic equations as explained in more detail in Szabo (2014d). For water, where $\gamma = 2$, $\beta_E(f) = 0.0$. The underlying loss mechanisms originate with atomic and molecular interactions, and modeling leads to the study of differential equations with fractional derivatives. A formal treatment of absorption mechanisms can be found in Holm (2019).

3.5.2 Absorption in the time domain

Of course, there must be an equivalent way of looking at absorption in the time domain. This is done through what is called a "material impulse response function," mirf,

$$\mathrm{mirf}(t, z) = \mathfrak{J}_{-i}^{-1}\{\exp[\gamma_T(f)z]\}.\tag{3.23}$$

For a pulse at any distance, z, the overall shape can be expressed as a convolution of the starting pulse shape, $p_0(t)$, and the material impulse response function,

$$p(t, z) = p_0(t) *_t \mathrm{mirf}(t, z),\tag{3.24}$$

which is the equivalent of the frequency block multiplication,

$$p(f, z) = P_0(f)MTF(f, z).\tag{3.25}$$

Examples of pulses at three different distances are given in Fig. 3.11. In this case, the initial pulse, $p_0(t)$, was a symmetric Gaussian. Note with increasing distance, the three propagation terms from Eq. (3.21) affect the pulses: first, the pulses are delayed by the distance-related travel times, second, the absorption reduces the pressure-pulse amplitudes with increasing distance, and finally, third, the dispersion term distorts the pulse shape with a spreading asymmetry. The peak pressure amplitude drops off roughly as $(a_0 z)^{1/\gamma}$ (Szabo, 2014d).

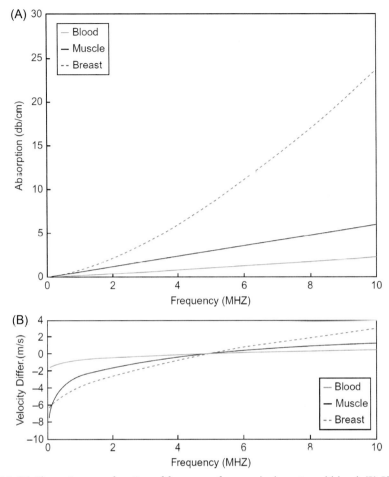

Figure 3.10 (A) Absorption as a function of frequency for muscle, breast, and blood. (B) Phase velocity dispersion difference minus a midband (5 MHz) sound speed value for the same tissues. *From Szabo, T. L. (2014d). Attenuation. In* Diagnostic ultrasound imaging: Inside out *(2nd ed.). Elsevier.*

3.5.3 Absorption Filter Simulator

Examples of absorption in action can be seen by using the Absorption Filter Simulator shown in Fig. 3.12. Similar to the Fourier Filter Simulator, different input pulse shapes can be selected. Then, for the absorption filter which calculates the MTF according to Eq. (3.20), a dropdown menu with different materials is offered from which to select. There is also an option to define a custom material; sound speed and absorption for a range of materials can be found in Appendix D. This selection in combination with the distance slider sets up the calculation of what the waveform looks like after it has propagated for the selected distance in the desired material. Note by choosing an

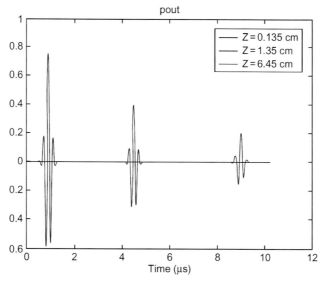

Figure 3.11 Changes in pressure−pulse shape of an initially Gaussian pulse propagating in a medium with a 1 dB/MHz/cm absorption with $y = 1.5$ for three different increasing propagation distances (z). *From Szabo, T. L. (2014d). Attenuation. In* Diagnostic ultrasound imaging: Inside out *(2nd ed.). Elsevier.*

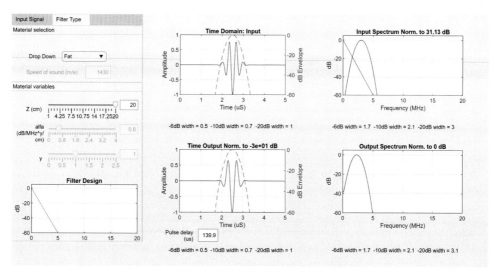

Figure 3.12 The Absorption Filter Simulator's input panel allows the selection of the input pulse and material. An manual override option permits independent input of α_0, y, and z. The output panel on the right provides the input and output pulses and spectra.

impulse response input, the *mirf* can be viewed directly. A user selectable mode permits the user to freely select reasonable values of absorption coefficient and power exponent γ. The top output panel gives the input time waveform, its spectrum and MTF(f); the bottom shows the output pulse and its spectrum. An example for a 3 MHz Gaussian pulse (0.1 variance) propagating 20 cm into fat is shown on the graphical user interface (GUI).

3.6 Lab 3: exploration of signals, filters networks and imaging of thin materials

3.6.1 Signal exercises with simulator applications

This chapter introduces a few signals typically used in imaging systems and explores their properties using the temporal and spectral domains. As discussed in Section 3.2, the Fourier transform links the two domains through well-known integral relationships (Eqs. 3.1 and 3.2). The exercises with the **Fourier Transform Simulator** (Fig. 3.2) permit the student to easily visualize the properties of the transform by presenting transform pairs for many types of signals without needing to manipulate equations.

As we will see in Chapter 4, transducers are resonant structures, so the signals used to drive them are often impulses and tone bursts. The simulator provides a basic signal generator with a number of adjustable parameters to set the signal type and central frequency, and windowing function properties. The student gains an intuitive appreciation for such signals and the inverse relationship between temporal extent of the signal, or burst duration, and the spectral bandwidth, a fundamental property of the Fourier Transform. This property is especially important for understanding that a short duration signal achieved by windowing a single frequency tone requires a significant amount of bandwidth to properly represent it. Conversely, narrowing the bandwidth requires extending the burst duration. The student is also introduced to the existence of negative frequencies, though the frequency plots can be restricted to the positive spectrum if desired. Impulses and rectangular signals, both unipolar and bipolar, are also provided, and the student is encouraged to explore the relationship between pulse width and bandwidth.

The idea that the imaging system can be represented by a series of functional blocks that transform the signal as it passes from block to block naturally leads to the concept of "filters" that modify the data by convolution of the signal with the filter. Signal processing, that is, the manipulation of signals, is naturally introduced as a way to either model physical processes that transform signals, or mathematical procedures that can be applied to signals to modify them according to the desired application. Convolution with a filter in one domain is multiplication by the transform in the other domain, providing another easily remembered method of intuitively constructing the Fourier transform of a signal. The idea that windowing of a continuous tone represents multiplication by an envelope function is fundamental to understanding how to obtain the Fourier transform of the product using a few easily remembered transform pairs and develops an understanding of the effect of convolution.

These concepts are reinforced through the use of the **Fourier Filter Simulator** (Section 3.3.2) which builds on the framework of the Fourier Transform Simulator to offer a selection of filter types and adjustment parameters such as bandwidth, and filter cutoff slope. With these options, the student can observe their impact on sidelobe levels and spectral bandwidth, for a select number of signals with which they are already familiar. Because the simulators provide numerically accurate plots, students can obtain quantitative measures and develop a practical sense of the relevant concepts and their impact on design requirements for various component blocks of the imaging system.

The imaging system block diagram contains a mix of physical and electronic functions, and the transducer, in particular, can be well modeled using electrical circuit analogs. In preparation for this representation discussed in the next chapter, modeling cascaded linear systems is introduced with the *ABCD* network matrix, a powerful and convenient computational tool. While easily implemented in modern programming languages such as MATLAB®, the ***ABCD* Simulator** provides a graphical approach to building up a network of passive linear circuit elements that can be used to model many types of filters, and particularly transducers. The exercises provide an introduction to manipulating network elements to observe first- and second-order networks and their solutions. Furthermore, the concept of input impedance of a network is presented and applied to impedance matching layers. In Chapter 4, these skills will be used to model specific transducers such as the one used in the imaging labs, and stacks of thin materials.

Finally, with the introduction to absorption, the **Absorption Filter Simulator** (Section 3.5.3) permits simulation of the effects of attenuation on a waveform as it propagates through a medium. On the left tab, an input pulse shape can be selected. Then a dropdown menu allows visualization of the effect of absorption of different materials on pulse shape for different propagation distances.

3.6.2 Experiments and exercises with the Vantage™ system

The hands-on laboratory uses the arbitrary waveform capability of the Vantage to transmit several typical types of signals through the transducer. These include unipolar pulses, bipolar wavelets, tone bursts, frequency sweeps, and Gaussian windowed bursts. The drive waveforms can be plotted and measured using the smooth "mirror" interface of the acrylic block used in Chapter 2. Representative plots are shown in Fig. 3.13.

Comparing panels B and C in Fig. 3.13, one can see that the measured waveform in C is slightly different from the simulated two-way waveform in B because, in the simulated field, the transducer impulse response is approximated.

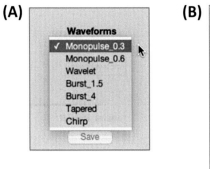

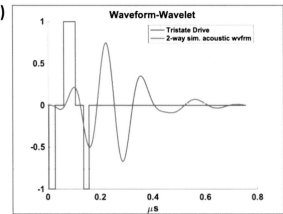

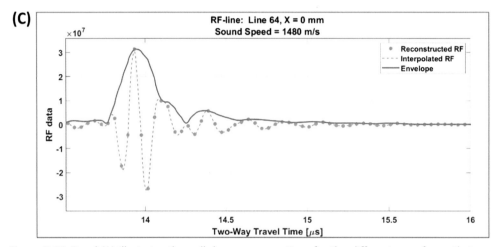

Figure 3.13 Panel (A) illustrates the pull-down menu options for the different waveforms that can be transmitted; "Monopulse" is selected. Panel (B) is a plot of the "Wavelet" type waveform, including the tristate drive (blue line) and the two-way acoustic waveform (red line). The bottom Panel (C) presents the A-line data collected using the acrylic block.

In these experiments, the various waveforms are then used to image the tissue mimicking phantom so that the student can see the changes in appearance of the various targets. The tradeoffs in intensity and resolution are apparent, as illustrated in Fig. 3.14 when using a short impulse and a longer chirp burst.

Finally, a number of thin materials are imaged using the different waveforms to illustrate how reverberation from such materials depends on frequency and waveform type. Utilizing the Layered Medium Simulator from Chapter 2, the students are able

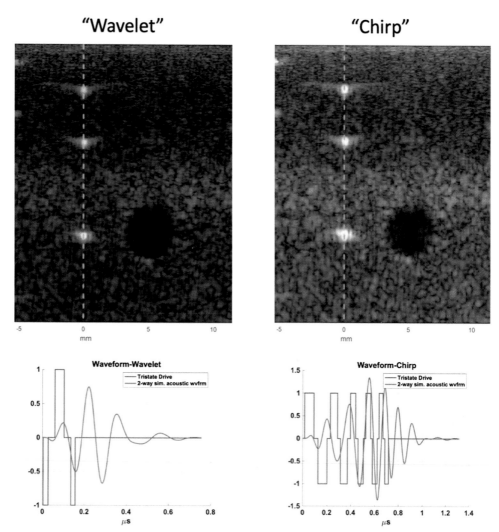

Figure 3.14 The left panel presents imaging using the Wavelet signal, and the right panel presents the image using a Chirp signal (frequency sweep). The image gain settings are the same, and one sees that the chirp image is brighter and blurrier.

to predict the reflection coefficients that arise when using thin sheet materials such as glass, plastic, metal, and graphite, for which the bulk speed of sound, impedance, and thickness is known. For example, several materials of interest are illustrated in Fig. 3.15A. An image of the reverberations from a glass slide is evident in Fig. 3.15B reminiscent of Section 2.5 and the Layer Pulse Simulator. The initial pulse echo from the glass slide and its multiple reflected echoes can be seen in Fig. 3.15C.

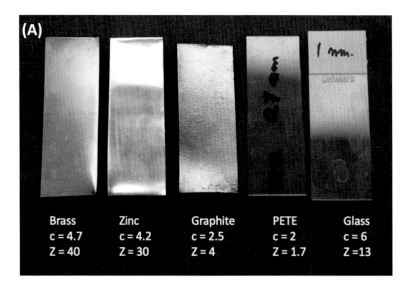

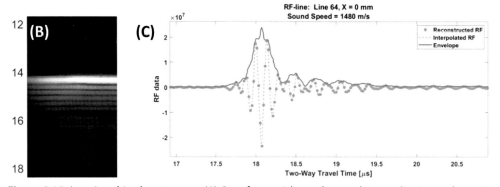

Figure 3.15 Imaging thin sheet targets. (A) Set of materials used to explore qualitative and quantitative backscattering and imaging properties. (B) Image of a glass microscope slide (1 mm thick) shows a series of reverberation echoes. (C) A linear plot of the RF line at the interface permits quantification of the normal reflection coefficient and of the reverberations.

References

Bracewell, R. N. (2014). *Groundwork. The Fourier Transform and its Applications* (3rd ed., pp. 5−6). India: McGraw Hill Education.

Holm, Sverre (2019). *Absorption Mechanisms and Physical Constraints. Waves with Power-Law Attenuation* (pp. 95−116). Springer.

Szabo, T. L. (2014a). *Overview. Diagnostic ultrasound imaging: Inside out* (2nd ed.). Elsevier.

Szabo, T. L. (2014b). *The Fourier transform. Diagnostic ultrasound imaging: Inside out* (2nd ed.). Elsevier.

Szabo, T. L. (2014c). *Acoustic wave propagation. Diagnostic ultrasound imaging: Inside out* (2nd ed.). Elsevier.

Szabo, T. L. (2014d). *Attenuation. Diagnostic ultrasound imaging: Inside out* (2nd ed.). Elsevier.

CHAPTER 4

Transducers

4.1 Overview

4.1.1 Transducer: the most critical part of an ultrasound system

A basic functional diagram for an imaging system is given by Fig. 4.1. The electrical signals are shown in gray. An electrical waveform is created in the first block E. This signal eventually enters the transducer shown in green which as a transmitter block, G_T, converts the waveform into an acoustic pressure signal. The pressure travels into an absorbing medium and returns to the transducer as a pressure pulse echo signal for reception, G_R, where it is reconverted into an electrical waveform.

Here a single element transducer will be emphasized. For imaging or other applications such as high-intensity focused ultrasound (HIFU), an array, consisting of many small transducers operating together, is required. All the transduction principles in this chapter still apply with two important differences: a transducer element in an array is physically smaller and the signal entering each individually addressable element is usually slightly different. Small, single element 5 MHz transducers are shown in the left side of Fig. 4.2; on the right is a rather large 0.6 MHz HIFU spherically focusing array consisting of 170 elements in an array. The wires that connect to the individual array elements are shown along with part of the electronic HIFU array hardware.

4.1.2 In this chapter you will learn

Why transducers are the critical component of any ultrasound system is explained. A transducer is an electromechanical device which operates both as a transmitter and receiver of pulse echo signals. Transducer functionality can be modeled as an equivalent circuit by the Transducer Simulator. Through this simulator, the unique hybrid operation of a transducer, operating in both acoustic and electrical realms, is demonstrated. Both electrical and acoustical parts of the transducer can be characterized and modified to improve its design and efficiency. Combinations of input waveforms and transducer responses can be selected to optimize imaging response or to operate at multiple frequencies.

4.2 Introduction to transducers and equivalent circuits

4.2.1 What is a transducer?

In Chapter 1, Sections 1.6.1 and 1.6.2, a simple A-line pulse echo system and a basic ultrasound imaging system with an array were explained. In both cases either a single transducer

Essentials of Ultrasound Imaging
DOI: https://doi.org/10.1016/B978-0-323-95371-9.00005-9

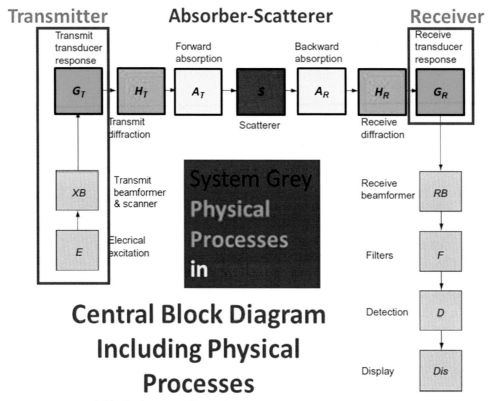

Figure 4.1 Central block diagram of an ultrasound imaging system highlighting transducers (Szabo, 2014a). *Adapted from Szabo, T. L. (2014a). Overview. In* Diagnostic Ultrasound Imaging: Inside Out *(2nd ed.). Elsevier.*

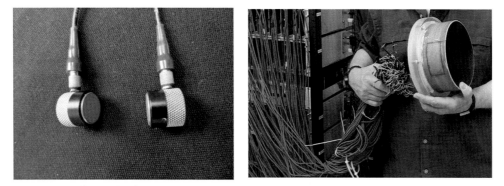

Figure 4.2 (Left) Single element 5 MHz transducer. (Right) 170 element 0.6 MHz spherically focusing array with connecting wires and system in background. Array used in (Chitnis et al., 2008).

or an array element functions as a reciprocal electromechanical device which converts an electrical drive pulse into a pressure pulse and converts it back to an electrical signal on receive. On the electrical end, a cable connects the transducer to the necessary electronic drive and receive circuitry. On the acoustic side of a transducer, usually identifiable by a protective coating or lens, some means of coupling the transducer to the medium or material or body is needed. A transducer, if it is so designed, can be immersed in water which conducts the sound to an immersed target material or it can be coupled mechanically to the object or body by a water-based gel or other type of couplant so that the transducer pressure waves can mechanically transfer to the target material. Obviously, without a transducer of some kind, there will be no acoustic waves in an ultrasound system.

While there are many mechanisms for converting electrical signals into acoustic pressures, the method most used for ultrasound is piezoelectric conversion. This kind of transducer is made of a special crystal, a material in which the electrical domains align preferentially in a given direction naturally or in which this alignment is achieved by poling, the one-time application of a high direct current field across its electrodes. This piezoelectric property is analogous to ferromagnetism, in which small magnetic domains within the metal can be aligned to magnetize the bulk along a particular direction. This is done by applying a very strong external magnetic field to the material, usually above the Curie temperature, a threshold above which the domains are mobile. If the material is cooled while under the external magnetization field, the alignment of small domains is frozen into the structure resulting a permanent magnetization. In a piezoceramic material, the Curie point is on the order of several hundred degree Celsius. A high voltage DC poling potential is applied to the electrodes when the material is near or above the Curie point, and released when cooled well below the Curie temperature, freezing in some portion of the applied electrical bias. Historically, the piezoelectric effect converts pressure to voltage (receive), and the inverse piezoelectric effect converts voltage to pressure (transmit) (Duck, 2022).

The basic principle is shown in Fig. 4.3. The crystal material is also nonconductive, so when electrodes are placed on opposite faces, the transducer is essentially a capacitor with special transduction properties. Piezoelectric material is dielectric; it has a clamped capacitance with a thickness d and an area A,

$$C_0 = \frac{\varepsilon^S A}{d},\tag{4.1}$$

in which ε^S is a clamped dielectric constant under the condition of zero strain.

For elastic solids (Szabo, 2014c), terms analogous to pressure and particle velocity are stress, T, and strain, S, and they are related by Hooke's law, $T = C^D S$, where C^D is an elastic constant. Because of piezoelectricity, the Hooke's law for this capacitor has an extra term,

$$T = C^D S - hD\tag{4.2}$$

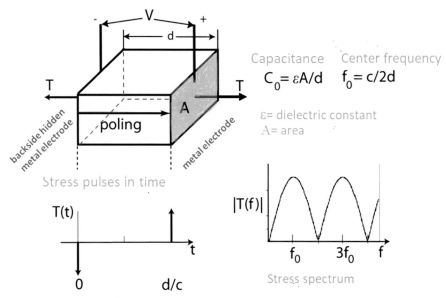

Figure 4.3 (Top) Diagram for a thickness expander piezoelectric crystal radiating into a medium matched to its impedance. (Bottom left) Stress time response. (Bottom right) Stress frequency response. *Adapted from Szabo, T. L. (2014b). Transducers. In* Diagnostic Ultrasound Imaging: Inside Out *(2nd ed.). Elsevier.*

in which h is a piezoelectric constant. The dielectric displacement D is given by

$$D = \varepsilon^S E = \frac{\varepsilon^S A V}{dA} = \frac{C_0 V}{A}. \tag{4.3}$$

where E is the electric field, and V is applied voltage. The speed of sound between the electrodes is given by $c = \sqrt{C^D/\rho}$, where the elastic stiffness constant C^D is obtained under a constant dielectric displacement D. With reference to Fig. 4.3, if a voltage impulse is applied across the electrodes, the inverse piezoelectric effect creates forces on the left and right sides of the transducer, given by

$$F(t) = TA = (hC_0V/2)[-\delta(t) + \delta(t - d/c)], \tag{4.4}$$

for which we have assumed that the media outside the electrodes has the same acoustic impedance as the transducer (see Fig. 4.3). To obtain the spectrum of this response, take the Fourier transform of Eq. (4.4),

$$F(f) = -i(hC_0V)e^{-i\pi fd/c}\sin[\pi(2n + 1)f/2f_0], \tag{4.5}$$

an expression with maxima at odd harmonics (note that $n = 1, 2, 3$, etc.) of the fundamental resonance $f_0 = c/2d$ (shown in Fig. 4.3 as the stress spectrum). Thus we observe that the piezoelectric transducer is a resonant structure of half-wavelength thickness at the fundamental resonant frequency.

4.2.2 Three-port transducer model

Expanding the concept shown in Fig. 4.3, the piezoelectric element can be viewed as a three-port device as illustrated by Fig. 4.4. An electrical signal is applied through an electrical port (across electrodes) and, from the inverse piezoelectric effect, acoustic pulses appear on the acoustic ports on either side. Physically, the crystal expands and contracts laterally and moves the electrode faces along what we call the propagation axis z.

This representation is a simplification. The equivalent circuit model which is hidden inside the box is an arrangement of $ABCD$ matrices not shown here but they are explained in Szabo (2014c, 2014d). These details need not concern us because we have the functional equivalent of this three-port model which forms the heart of the soon to be explained Transducer Simulator. What is important is understanding how attachments can be made to the ports.

In order to make the transducer useful, we must find a means to efficiently couple a suitably shaped electrical signal into port 3 and maximize and shape a pressure pulse emerging from acoustic port 1. Fig. 4.5 is a diagram of the structure of a typical ultrasound transducer.

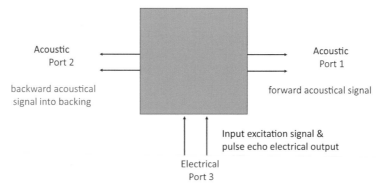

Figure 4.4 Piezoelectric three-port model.

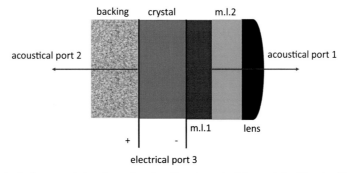

Figure 4.5 Typical ultrasound structure showing the piezoelectric crystal with a backing on acoustical port 2 and two matching layers and a lens on acoustical port 1.

Fig. 4.5 represents the layered physical structure of a transducer. On either side of the crystal are the acoustic ports. At acoustic port 1 on the right, matching layers efficiently couple acoustic energy from the crystal which typically has a high acoustic impedance to tissue which has a low one. The red line on the right indicates the propagation or $+z$-axis. The purpose of the backing, along the $-z$-axis, is prevent the port 2 signal from interfering with the intended port 1 signal and to broaden bandwidth. Opportunities to explore the effects of these "attachments" will be treated in more detail with the Transducer Simulator. This picture (Fig. 4.5) is still incomplete. It is possible to connect a network on the electrical port too! This can be done with an $ABCD$ tuning network as explained shortly.

The one-dimensional equivalent circuit transducer behind the simulator works remarkably well in predicting the behavior of a transducer in real practical terms as will be shown in the next sections. This is a conceptual model of considerable practical value.

4.2.3 Transducer blocks

Before investigating the workings of a transducer more deeply, we return to the central block diagram in Fig. 4.1. Here we see the essential parts of the transducer functional blocks: E, the electrical signal source, the transmit part of the transducer, G_T, the scatterer, S, which in this case will be a mirror once again, and the receive part of the transducer, G_R. Physically, the receive part of the transducer is the same as the transmit part and that is because piezoelectricity is reciprocal. In other words, a pressure pulse coming into the transducer's acoustic port creates a voltage on the electrical port. Once again, there is an electrical source feeding the transducer by the inverse piezoelectric effect, then the transmit part of the transducer creates a pressure wave which hits a scatterer and returns back to the transducer and is reconverted to an electrical signal through the piezoelectric effect.

In order to represent these steps in more detail to see how they work together to create a pulse echo signal, we refer to Fig. 4.6. On the left is the electrical source E and its source impedance connected to an $ABCD$ block representing an electrical

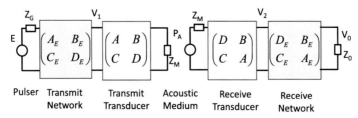

Figure 4.6 Expanded blocks from the central block diagram: E, G_T, S, and G_R.

matching network. There is an intermediate voltage V_1 across the electrical impedance of the transducer. We can also take the three-port model and turn it into a two-port *ABCD* model with the electrical port 3 on one side and acoustic port 1 on the other side. In this case, the other acoustic port 2 is absorbed into the model but need not be shown separately. The one-way transduction process ends in a pressure, P_A across a material load, Z_M. For the pulse echo or round-trip case, the pressure is scattered and returns to the transducer and reenters the transducer now operating in its receive mode. Since for receive transduction, the same transducer is used in a reciprocal function, we find the matching network and transducer receive mode to be simply the same matrix elements rearranged. Again, the receive voltage across the transducer is V_2. Finally, there is a final output voltage, V_0, across an output electrical load. The final two-way transfer function is the product of individual transfer functions from the blocks in Fig. 4.6:

$$\frac{V_0}{E} = \frac{V_1}{E}\frac{P_A}{V_1}\frac{V_2}{P_A}\frac{V_0}{V_2}. \tag{4.6}$$

A complete transducer equivalent circuit, constructed of *ABCD* blocks, can be found in Szabo (2014c) and it is used in the Transducer Simulator.

4.2.4 Electrical transducer port

The electrical impedance looking into the electrical port 3, called the input impedance, is a complex function of frequency which changes with different designs and attachments. Certain features are recognizable and nameable as shown in Fig. 4.7. Recall that the transducer is also a capacitor, so we expect to see a capacitive reactance term, $-i/(2\pi f C_0)$. In addition, because the transducer converts an electrical signal into an acoustic pressure over a certain frequency band, indirect electrical evidence of that acoustic activity shows up as an acoustic impedance. This impedance, Z_A, has a real part, called the radiation resistance, $R_A(f)$ and an imaginary reactance, $X_A(f)$, so that the total transducer impedance, $Z_T(f)$, at port 3 is

$$Z_T = Z_A - i(1/\omega C_0) = R_A(f) + i[X_A(f) - 1/\omega C_0]. \tag{4.7}$$

In general, the radiation resistance is positive and roughly centered at the center frequency, f_0,

$$R_A(f_0) = R_{A0} = \frac{2k_T^2}{\pi^2 f_0 C_0}\left(\frac{Z_C}{Z_L + Z_R}\right) = \frac{2k_T^2}{\pi^2 f_0 C_0}\left(\frac{Z_C}{Z_B + Z_w}\right), \tag{4.8}$$

where k_T is the piezoelectric coupling constant, Z_C is the crystal impedance, Z_L is the acoustic impedance looking to the left of acoustic port 1, and Z_R is the impedance to the right of port 2. If there are no matching layers, typically $Z_L = Z_B$, the

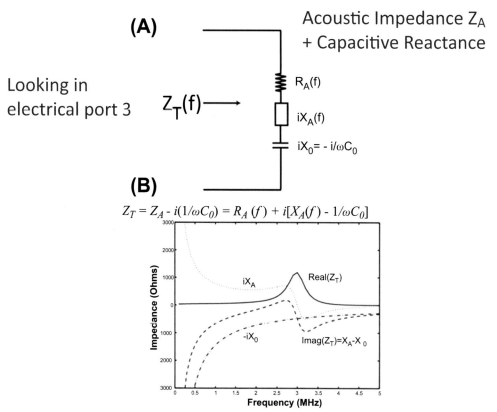

Figure 4.7 (A) transducer equivalent circuit model with radiation resistance, real($Z_T(f)$) = $R_A(f)$, radiation reactance, $X_A(f)$, and capacitive reactance, $X_0(f)$. (B) Transducer impedances are a complicated function of frequency. In this case, $f_0 = 3$ MHz. *From Szabo, T. L. (2014b). Transducers. In Diagnostic Ultrasound Imaging: Inside Out (2nd ed.). Elsevier.*

backing impedance, and $Z_R = Z_W$, the water or tissue impedance for a medical application. The acoustic reactance is both positive and negative and goes through zero at f_0.

4.2.5 Transducer Simulator

The Transducer Simulator provides a generic transducer equivalent circuit model with several degrees of freedom. General controls are available such as selecting the center frequency of the transducer. The simulator operates in two basic output modes: time/frequency and impedance/loss: these are controlled by a select plotting toggle switch. Examples of each mode can be seen in Figs. 4.8 and 4.9. In addition, one-way or round-trip (pulse echo) options are available by a pulse switch. In the impedance plot option, Fig. 4.8, the real and imaginary parts of the transducer impedance are shown. Below the impedance plots are

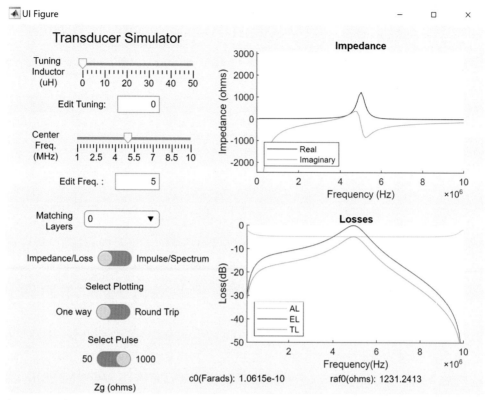

Figure 4.8 The Transducer Simulator in the impedance/loss mode for a 5 MHz transducer with no matching layer.

acoustic loss, AL(f), and electric loss, EL(f), curves on a dB scale. Note the sum of the curves is the transducer loss, TL(f), often expressed in dB and in linear terms,

$$TL(f) = AL(f)EL(f). \tag{4.9}$$

Specific controls for electrical matching and matching layers which affect these loss curves will be dealt within the next sections. In the time–frequency mode, Fig. 4.9, the transducer impulse response signal and its envelope, $e(t)$, and spectral magnitude, $|E(f)|$, are displayed.

4.2.6 Acoustical loss

The electrical power reaching R_A is converted to acoustic power and then divided proportionally between the left and right acoustic ports, depending on their loading. This split can be quantified by the acoustical loss (Szabo, 2014b) which at the center frequency, f_0, is simply the acoustic impedances looking left and right, which in this case would be the backing and water (tissue) loads,

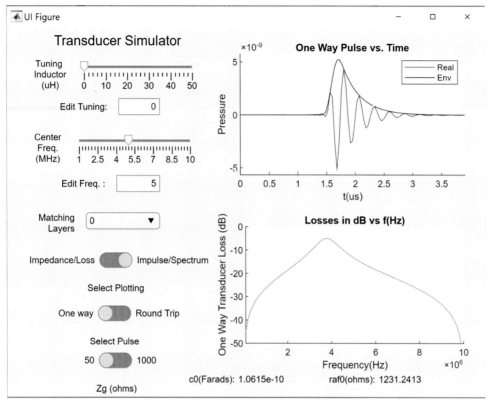

Figure 4.9 The Transducer Simulator in the impulse response/spectrum mode for a 5 MHz transducer with no matching layer.

$$AL(f_0) = \frac{Z_R}{Z_R + Z_L} = \frac{Z_W}{Z_B + Z_W}, \qquad (4.10)$$

which for $Z_W = 1.5$ and $Z_B = 3$ MRayls results in AL $= 0.33$ (-4.82 dB). If a matching layer was used to match the impedance of the crystal $Z_C = 34.3$ MRayls (Eq. 2.24); then the apparent impedance looking right toward the tissue would equal Z_C so that AL(f_0) becomes 0.92 (-0.36 dB), an improvement of $+4.5$ dB. A comparison of the AL curves at f_0, for Figs. 4.8 and 4.10, shows that without the layer, AL $= -4.8$ dB; and with it, AL $= -0.4$ dB or a gain of $+4.4$ dB.

4.2.7 Electrical loss

In order to improve transducer design, it is necessary to minimize losses. The overall one-way transfer function from electrical port 3 to acoustic port 1 is called transducer loss, TL(f) from Eq. (4.9). From Figs. 4.6 and 4.7, it is possible to write an equation for the electrical loss (Szabo, 2014b), EL(f). To find the power delivered to the real

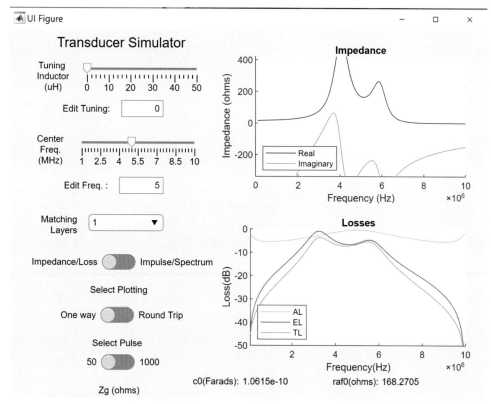

Figure 4.10 The Transducer Simulator in the impedance/loss mode for a 5 MHz transducer with one matching layer and $R_G = 50$ ohms.

part of the transducer impedance $R_A(f)$, include the source, a real source impedance $Z_G = R_G$, the matching network. The transducer impedance, $Z_T(f)$, and A_E and B_E are equivalent network parameters from the *ABCD* matrix.

$$EL = \frac{4R_A R_G}{|A_E Z_T + B_E|^2}. \tag{4.11}$$

To compensate for the capacitance in the circuit (see Fig. 4.7A), a series inductor is often added to the matching network. As derived in more detail (Szabo, 2014b), Eq. (4.11) becomes, with a series inductor, L_S, from the Fig. 3.6 *ABCD* matrix choices and Z_T from Eq. (4.7),

$$EL = \frac{4R_A R_G}{(R_A + R_G)^2 + (X_A - 1/\omega C_0 + \omega L_S)^2}. \tag{4.12}$$

Near $f_0 = 5$ MHz, the capacitive reactance can be canceled out by the inductor value, $L_s = 1/(\omega_0^2 C_0)$, leaving only real terms in Eq. (4.12) and reducing the electrical loss. This process is often called "tuning out" the capacitive reactance. Occasionally, to broaden the frequency response of the transducer, the tuning is performed for a frequency value that is slightly different than the natural resonant frequency. One can observe the effect of off-resonance tuning on the impedance and loss spectra by adjusting the value of the tuning inductance in the Transducer Simulator. Usually, $R_G = 50$ ohms.

As a specific example, refer to Figs. 4.10 and 4.12 (and Figs. 4.11 and 4.13 for the corresponding time/frequency plots) for the case of a matching layer without and with a tuning inductor. From the panel, $R_A(f_0) = 168$ ohms and $C_0 = 106$ pf or a reactance of 299 ohms at f_0. Then the tuning inductor is $L_s = 9.56$ µH. Since $X_A(f_0) = 0$, Eq. (4.12) for the untuned case is $EL(f_0) = 0.0614$ or -12.1 dB. The inductor cancels out the imaginary part so that the tuned case is $EL(f_0) = 0.177$ or -7.5 dB, an increase

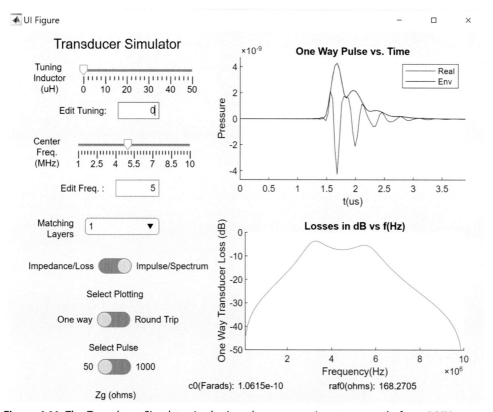

Figure 4.11 The Transducer Simulator in the impulse response/spectrum mode for a 5 MHz transducer with one matching layer and $R_G = 50$ ohms.

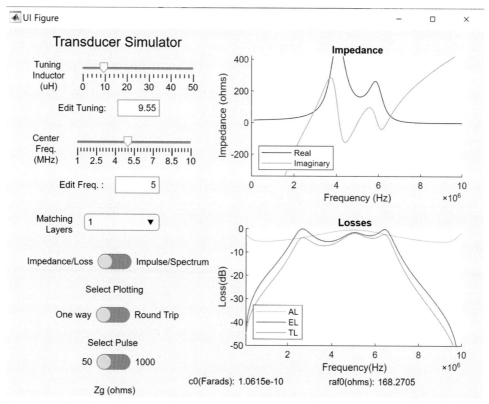

Figure 4.12 The Transducer Simulator in the impedance/loss mode for a 5 MHz transducer with one matching layer and tuning inductor and $R_G = 50$ ohms.

of $+4.6$ dB. Comparison of the EL curves at f_0, for Figs. 4.10 and 4.12, shows that without the inductor: $EL = -6.1$ dB; and with it, $EL = -1.5$ dB or a gain of $+4.6$ dB.

4.2.8 Insertion Loss

"Insertion Loss" is a measure of the round trip throughput of a transducer based on the assumption of a perfect reflector. Imagine the round-trip setup in Fig. 4.6 with everything between the source E and its source impedance, Z_G, and the output impedance, Z_0, removed. Then the power delivered to the load would be calculated from an equation similar to Eq. (4.12),

$$EL = \frac{4Z_0 R_G}{(Z_0 + R_G)^2}, \tag{4.13}$$

assuming Z_0 is real. Most often, $Z_0 = R_G$. Now imagine inserting everything back in Fig. 4.6 and recalculating the power to the load. The ratio of the power delivered with the inserted network to the power without the network is the insertion loss. For simplification, if everything is kept the same for receive (tuning load, etc.) then IL = TL^2(linear) or IL$_{dB}$ = 2TL$_{dB}$. This of course means that the benefits of matching layers and tuning are doubled in terms of dB.

4.2.9 Transducer design

While we have concentrated in improving efficiency especially at one frequency point, this concern is only part of overall transducer design. From Figs. 4.8, 4.11, and 4.13, it is evident that both efficiency and bandwidth have been improved substantially. The wider bandwidth has come at the expense of pulse envelope shape.

The sought after ideal in transducer design is not only to obtain greater efficiency resulting in larger echo amplitude but also to create a favorable impulse response

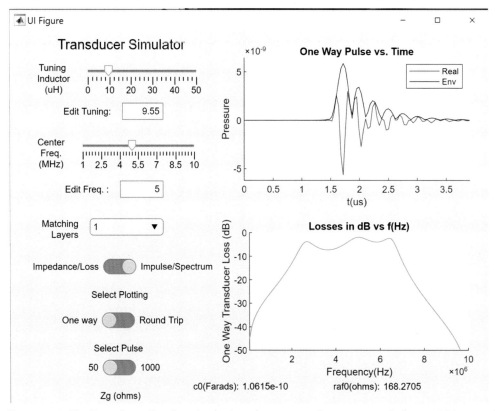

Figure 4.13 The Transducer Simulator in the impulse response/spectrum mode for a 5 MHz transducer with one matching layer and a tuning inductor and $R_G = 50$ ohms.

envelope shape. Beauty in transducer pulse design is a short, rounded pulse decaying rapidly with no trailing time sidelobes. For imaging, the envelope of the pulse is used, so if a target is hit by the pulse, it's apparent size or "time resolution" will be approximately the −6 dB length of the pulse, that is, the temporal width of the envelope between the −6 dB points. If it is an ideal pulse shape, then it will appear as one unambiguous blob (later in the course will be examples of this). The pulse in Fig. 4.9 is not bad insofar as shape is concerned but it is too long and has poor resolution. The main lobe in Fig. 4.11 is shorter but has a large trailing time sidelobe. Finally, in Fig. 4.13, the pulse has short main lobe but so many trailing time sidelobes that it might appear as a mother duck followed by her ducklings. Measures of how fast the pulse decays are the −10 and −20 dB outer pulse lengths as given in Fig. 4.14.

There is another solution if the transducer has sufficient bandwidth derived from the spectral magnitude of the Fourier transform of the impulse response. Practical measures of bandwidth are the absolute bandwidth: the difference between the upper and lower frequencies −6 dB down from the spectral maximum. The center frequency is mean (halfway point) between these frequencies. The fractional, or relative

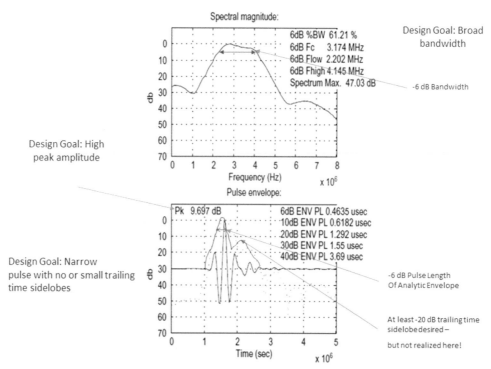

Figure 4.14 A transducer impulse response pulse and spectrum quantified against transducer design goals (Szabo, 2014e). *Adapted from Szabo, T. L. (2014e). Ultrasonic exposimetry and acoustic measurements. In Diagnostic Ultrasound Imaging: Inside Out (2nd ed.). Elsevier.*

bandwidth (in percent) is 100 times the ratio of absolute bandwidth to center frequency. It is still possible to shape the overall output pulse by changing the shape of the drive pulse and its placement in the transducer's useable bandwidth. This is the game that can be played with the Fourier Filter Simulator that includes a transducer response which can be combined with different input drive pulse shapes such as the one shown in Fig. 3.4.

Finally, the examples have shown that transducer design is complicated. A good example of transducer design (high sensitivity, a short pleasing pulse shape and adequate bandwidth) is included in the Transducer Simulator as "option 2." See Fig. 4.15. This is an approximation of the response of the transducer we will be using in our labs, the L11−5, a linear array spanning a frequency range from 5 to 11 MHz.

The transducer simulation is compared to data for this transducer in Fig. 4.15. The center frequency is $f_c = 7.1$ MHz as determined from the from the −6 dB bandwidth frequencies. The −6 dB bandwidth is 70%; the −10 dB is 87% and the −20 dB is 121%. For a period of $T = 1/f_c$, the −6 dB envelope width is 1.2 T, the −10 dB is 1.6 T and −20 dB width is 3.0 T.

4.2.10 Transducer applications

General transducer design principles serve a variety of transducer applications even though medical imaging transducers have been emphasized. Matching layers can be adapted to different loads such as steel and plastics, foods, and earth, and for long-range sonar and short-range air targets. The primary form of medical imaging transducers is the array, a geometric arrangement of usually identical miniature transducers each of which most often have separate electrical connections. More about arrays will be found in Chapters 6−8. Medical imaging arrays are shaped and sized to operate in

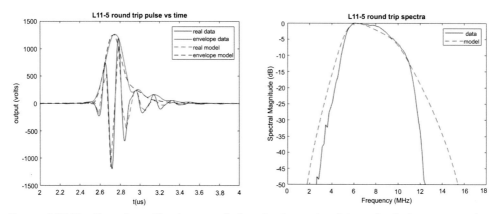

Figure 4.15 The Transducer Simulator predictions in the optional 2 mode design compared to round-trip data for the L11−5 array.

certain regions of the body for specific applications. Array construction is related to imaging format as discussed in Chapter 8. In general, transducer frequencies for imaging range from tens of Hertz to hundreds of kHz for oceanography and geophysics, 0.5−10 MHz for nondestructive evaluation, 1−30 MHz for medical imaging applications, and hundreds of MHz to several GHz for acoustic microscopy. In addition, transducers serve many other diverse applications ranging from cavitation cleaners, sonochemistry, ultrasonic machining, noncontact production line monitoring, flea collars and surgery. Even though piezoelectric-based transducers are the most common, other transduction principles such as those used in capacitive micromachined ultrasound transducers (CMUT) and piezoelectric micromachined ultrasound transducers (PMUT), electromagnetic, electrostatic, and acoustic radiation force are in use.

4.3 Lab 4: exploring transducer modeling and the acoustic stack

4.3.1 Concepts explored with simulator applications

Transducers are essential components of an ultrasound imaging system and are highly specialized physical devices that have both electrical and mechanical properties. At their core is a piezoelectric material that converts electrical voltage to mechanical deformation and vice versa. Practical transducers are layered devices (Fig. 4.5) with material properties and dimensions that are optimized to provide mechanical efficiency by optimally coupling displacement of the piezoelectric material interface to pressure waves in the acoustic medium. Fortunately, electromechanical transducer behavior can be effectively modeled using electrical equivalent circuit analogs of the mechanical features, as described in Section 4.2.3 and 4.2.4. The network element model based on *ABCD* parameter matrix multiplication (Section 3.4) is used to represent some of the functional blocks describing transduction in the now familiar imaging system block diagram, recalled in Fig. 4.1. As in prior chapters, software simulator applications are used to convey important concepts through interactive input parameter controls and graphical output results.

The Transducer Simulator (beginning in Section 4.2.5) models a transducer as a series electrical network, diagrammed in Fig. 4.7A, with a few adjustable parameters and several selectable output modes providing the student with opportunities to develop their intuition. The impedance/loss display mode is used to illustrate the resonant and relatively narrowband behavior of a transducer and the loss curves indicate that efficiency depends strongly on selection of a backing material impedance and absorption, mechanical impedance matching, carefully chosen materials in quarter-wave layers, and on the use of inductive tuning to minimize the role of reactive (capacitive) network elements that are integral to the piezomaterial itself and to cabling. Exercises require the student to collect quantitative data as they evaluate various combinations of transducer parameters, and ultimately, optimize the design for bandwidth or efficiency, or to find a good compromise between the two competing metrics. Students are shown how to use -6 and -20 dB levels in both time and frequency domains to estimate performance.

Similarly, the Impulse/Spectrum output mode indicates how the transducer's Impulse Response and spectrum varies with center frequency, inductive tuning, and physical matching layer designs. The response is presented in both temporal and spectral domains, representative of the Fourier transform relationships explored in Chapter 3. The one- and two-way responses are calculated and reinforce the understanding of transmit-only (one-way) and transmit-receive (two-way and pulse echo) experimental configurations.

As presented in Fig. 4.15, the Transducer Simulator is able to model the Verasonics L11−5v transducer with good fidelity. The students are able to create new transducer designs and examine the Input Impedance and Loss curves, and to quantify performance in terms of design goals described in Section 4.2.9.

4.3.2 Experiments and exercises with the Vantage system

The Vantage system is used to study layered structures similar to those used in transducer structures, following on from the imaging of single layers in laboratory experiments in Chapter 3. Now with various materials to build up a passive acoustic stack, the students can observe the change in reflection coefficient as a function of material and signal type, and frequency. Fig. 4.16 illustrates the four kinds of experiments that students can easily conduct with thin materials and laminations, simply by measuring the A-line normal reflection. Given the sound speed c, the acoustic impedance Z, and the thickness d, students can use the simulators to estimate the relative reflection coefficients for the single- and double-layer cases (A, C, and D), and compare with

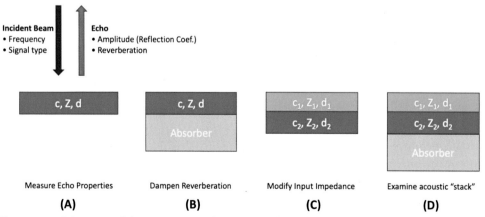

Figure 4.16 Schematic of the acoustic "stacks" used in the lab. Thin materials from Lab 3 are provided in various laminated configurations for students to use in making relative comparisons of reflection coefficient and reverberation length.

measurements. Reverberation signals between stacks (A, B, and D) can also be directly observed experimentally.

Thin layer materials with impedance values similar to that of piezoceramic (~30 MRayls) made of Zinc sheet metal represent the piezoceramic layer, while materials with smaller mechanical impedances (plastic, graphite) are used as matching layers. While not the same as a piezoceramic transducer, the glass microscope slide impedance can be "matched" to some degree by adding a quarter-wave plastic layer. The idea behind the experiments is that the effectiveness of the mechanical impedance match between a matching layer and a resonant layer will be quantifiable by measuring the reflection coefficient. As the students learned in Lab 3 by observing the reflection from a variety of materials, the reflection coefficient can be measured in a relative sense by observing the peak value of the A-line RF envelope at the interface of the matching layer compared with the reflection at the interface when the matching layer is absent.

The students can then use an absorbing backing material to dampen reverberations in the resonant layer. Adding the backing will significantly shorten the "ringing" of the response, which can be observed in the B-mode image as well as on the A-line plot. Using a short impulsive transmit signal, the students are able to measure an impulse response of sorts for their acoustic stack. Because laminations operating in the frequency range of the L11−5v are sensitive to details of their assembly, several material stacks have been prebonded and are provided for easy substitution to examine differences in refection coefficient and attenuation of reverberation, as a function of frequency over the L11−5v passband.

References

Chitnis, P. V., Barbone, P. E., & Cleveland, R. O. (2008). Customization of the acoustic field produced by a piezoelectric array through interelement delays. *Journal of the Acoustical Society of America*, *123*(6), 4174−4185.

Duck, F. (2022). Paul Langevin, U-boats, and ultrasonics. *Physics Today*, *75*(11), 42−48.

Szabo, T. L. (2014a). Overview. *Diagnostic ultrasound imaging: Inside out* (2nd ed.). Oxford: Elsevier, Chapter 2.

Szabo, T. L. (2014b). Transducers. *Diagnostic ultrasound imaging: Inside out* (2nd ed.). Oxford: Elsevier, Chapter 5.

Szabo, T. L. (2014c). Acoustic wave propagation. *Diagnostic ultrasound imaging: Inside out* (2nd ed.). Oxford: Elsevier, Chapter 3.

Szabo, T. L. (2014d). Development of one-dimensional KLM model based on ABCD matrices. *Diagnostic ultrasound imaging: Inside out* (2nd ed.). Oxford: Elsevier, Appendix C.

Szabo, T. L. (2014e). Ultrasonic exposimetry and acoustic measurements. *Diagnostic ultrasound imaging: Inside out* (2nd ed.). Oxford: Elsevier, Chapter 13.

CHAPTER 5

Beams and focusing

5.1 Overview

5.1.1 Diffraction

In terms of the block diagram shown in Fig. 5.1, there are two blocks for diffraction: one in transmit and another in receive. Based on the wave nature of the transmission of waves, diffraction adds more realism to wave propagation from a transducer. Diffraction is interesting because waves spill into regions not expected from geometric models, and they appear to bend around objects.

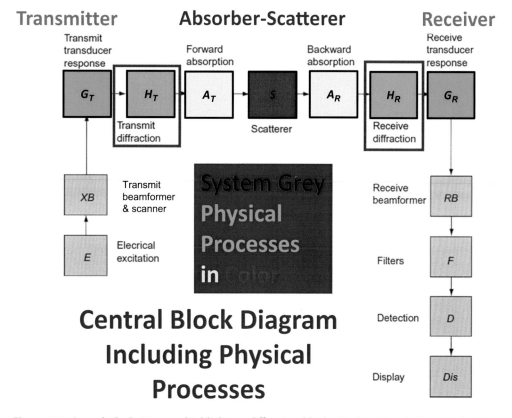

Figure 5.1 Central Block Diagram highlighting diffraction blocks (Szabo, 2014a). *From Szabo, T. L. (2014a). Overview. In Diagnostic Ultrasound Imaging: Inside Out (2nd ed.). Elsevier.*

Essentials of Ultrasound Imaging
DOI: https://doi.org/10.1016/B978-0-323-95371-9.00003-5

In Chapter 2 we encountered two representations of acoustic waves: k-rays representing sound traveling in straight lines and wavefronts expanding from different shaped sources. Now a third kind of representation of more realistic acoustic wave propagation will be presented. In this case, a slit or linear opening, called an "acoustic aperture (or window)," tens of wavelengths long, radiates waves in a surprising way. A comparison between k-rays and diffraction is shown in Fig. 5.2. On the left is a slit or aperture of length L shown in green. Sound pours through the slit in an upward direction as shown by the arrow. If the sound is in the form of rays (red arrows), that is, where the wavelength is very small with respect to the aperture, we would expect the intensity pattern of the field, shown in a gray scale representation, to be the same everywhere and to have the width L. On the right of Fig. 5.2 is another slit but this one is 40 wavelengths wide, and it produces a complicated pattern. These diffraction effects occur when the aperture is a few or tens of wavelengths wide. As explained in more detail later, diffraction is caused by the interference of waves coming from different parts of the aperture. This top-down representation of the field is called an intensity plot, shown again with labels and dimensions in Fig. 5.3.

Through our physics simulators, the relationship between the different parts of the beam will become clearer. The beam simulator also offers different representations of a beam. For example, the beam shown as an intensity plot in Fig. 5.3 is displayed as a pressure magnitude surface plot in Fig. 5.4.

5.1.2 How beams are formed

Recall that a spherical wavefront expands like an inflating balloon from a point source originally introduced in Section 2.3.1 by the Expanding Waves Simulator. In order to understand how beams are formed, an aperture or finite-length source of acoustic waves is needed. In a plane, a spherical wavefront is seen as a circle with an expanding radius

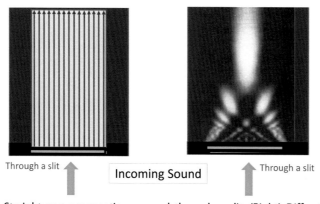

Figure 5.2 (Left) Straight rays propagating upward through a slit. (Right) Diffracted beam propagating upward through a slit.

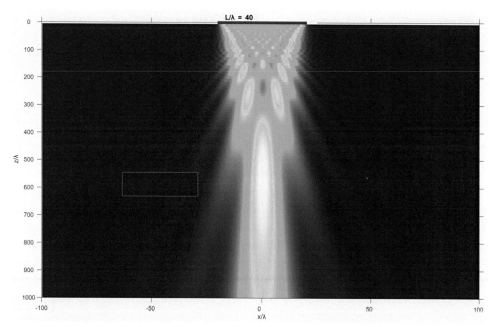

Figure 5.3 Intensity plot of an unfocused diffracted beam propagating downward along the *z*-axis from a 40 wavelength wide horizontal aperture (red) along *x* shown at the top of the plot.

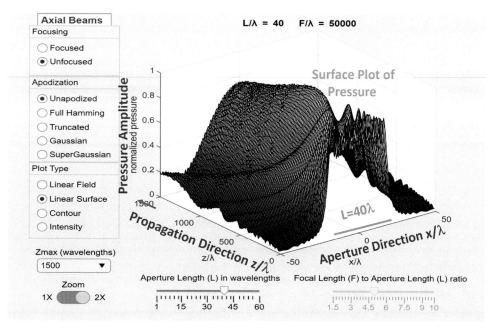

Figure 5.4 Surface pressure magnitude of a beam formed by a 40 wavelength wide aperture from the Axial Beams Simulator.

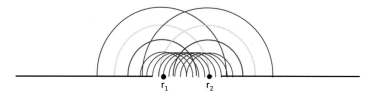

Figure 5.5 Open region between the horizontal lines is an aperture. Wavefronts from 9 point sources illustrate the spherical wavefronts at a specific time, t_1. Two of these are shown expanding at later snapshots in time.

with time. In Fig. 5.5, are 9 wavefronts from equally spaced point sources shown in blue at an instant of time. These wavefronts are clustered together forming a beam. In this figure, two of the wavefronts emanating from r_1 and r_2 are shown in three later snapshots of time. In the overlap region, the acoustic field is generated by the constructive interference of wavefronts. More about beam formation will be discussed in Chapter 6.

In order to compute fully developed acoustic wavefields like the ones in Figs. 5.3 and 5.4, more comprehensive models need to be developed. The Rayleigh integral approach is a diffraction model based on spherical wavefronts emanating from point sources in the aperture and will be described soon. First, however, an entirely equivalent diffraction model based on simple plane waves will be explained.

5.1.3 In this chapter you will learn

Unlike optics, in which the dimensions of lenses and apertures are more than a thousand times larger than a wavelength and waves can be approximated by straight rays, sources in ultrasound (transducers) have dimensions of just tens of wavelengths across, so that diffraction creates complicated wave patterns in the radiated acoustic field. These diffraction effects can be controlled to shape and focus the acoustic beam to increase resolution in certain selected regions of the field. Our main goal for this chapter is to discuss the nature of diffraction and explain how the continuous wave diffractive field can be predicted and controlled for design purposes. How an acoustic field can be quantified and characterized will be explained. In simple terms, a one-dimensional acoustic transducer aperture radiates a two-dimensional beam (with finite thickness in the third dimension) whereas a two-dimensional aperture radiates a three-dimensional beam. Interactive simulators provide immediate feedback on how changing the different degrees of freedom in design affects the shape of the beam.

5.2 Diffraction models for calculating beams

5.2.1 Spatial transforms

In order to calculate acoustic beams, we can use two- and three-dimensional k-rays which were introduced earlier in Section 2.6.2. Instead of point sources, consider elemental plane waves traveling in different directions **k** from a finite-length aperture as presented in Fig. 5.6. Each **k** vector has a magnitude $k = |\mathbf{k}|$ and is defined at an angle θ from the k_3 axis as depicted in a two-dimensional representation (in a plane) in Fig. 5.7A.

Now they are used for k-waves which are plane waves traveling in the vector direction of a k-ray. For a plane wave propagating along z, for example,

$$\exp[i(\omega t - \mathbf{k}\cdot\mathbf{z})] = \exp\left[i2\pi(ft - \tilde{f}z\cos\theta)\right], \tag{5.1}$$

where the plane wave can be interpreted in terms of temporal frequency f and spatial frequency $\tilde{f} = |\mathbf{k}|/2\pi$. Recall that a time waveform can be decomposed into frequencies or a spectrum. Similarly, a general propagating field with a given temporal frequency can be decomposed as a weighted sum over plane waves, or equivalently over spatial frequencies, $\tilde{f}_i$, that result from propagation in different directions (Fig. 5.7). Thus we can write a basic spatial transform as a plus i Fourier transform, $\mathfrak{I}_{+i}$, integral in terms, for example, of waves along x,

elemental plane waves

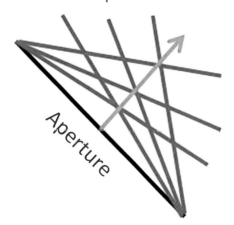

Figure 5.6 Elemental plane waves radiating in different directions from a line aperture.

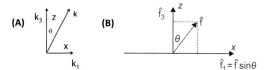

Figure 5.7 (A) **k** vector at an angle θ from the **k_3**-axis and has projections k_1 along x and k_3 along z. (B) Spatial frequencies corresponding to k-ray. *Adapted from Szabo, T. L. (2014a). Overview. In Diagnostic ultrasound imaging: Inside out (2nd ed.). Elsevier.*

$$G(\tilde{f}) = \Im_{+i}[g(x)] = \int_{-\infty}^{\infty} g(x)e^{i2\pi\tilde{f}\sin\theta x}\,dx, \tag{5.2}$$

and its inverse,

$$g(x) = \Im_{+i}^{-1}[G(\tilde{f})] = \int_{-\infty}^{\infty} G(\tilde{f})e^{-i2\pi\tilde{f}\sin\theta x}\,d\tilde{f}. \tag{5.3}$$

Much as the Fourier transform connects the time domain to the frequency domain, these spatial Fourier transforms connect the spatial domain (length coordinates) with the spatial frequency domain (wavenumber coordinates), where the latter is often called the k-space. We can construct an acoustic field from these elemental plane waves, each associated with a **k** vector, radiating from a finite length aperture as illustrated in Fig. 5.7.

The spatial frequency components can be combined, where the spatial frequency directions 1, 2, and 3 are associated with the cartesian coordinate directions x, y, and z, respectively,

$$\tilde{f}_1^2 + \tilde{f}_2^2 + \tilde{f}_3^2 = \frac{k^2}{4\pi^2} = \frac{f^2}{c^2} = \frac{1}{\lambda^2}$$

$$\tilde{f}_3^2 = \pm\left(\tilde{f}^2 - \tilde{f}_1^2 - \tilde{f}_2^2\right)^{1/2} \quad \text{if} \quad \tilde{f}^2 > \tilde{f}_1^2 + \tilde{f}_2^2 \tag{5.4}$$

$$\tilde{f}_3^2 = i\left(\tilde{f}^2 - \tilde{f}_2^2 - \tilde{f}^2\right)^{1/2} \quad \text{if} \quad \tilde{f}^2 < \tilde{f}_1^2 + \tilde{f}_2^2$$

in which $\tilde{f}_1$ is $\tilde{f}\sin\theta$ along x, etc. Note there is a correspondence to a previously used variable $u = \sin\theta$. These variables correspond to a spatial coordinate system and spatial frequency/k-space representation illustrated in Fig. 5.8 and Table 5.1.

To recap, the pressure at a field point is expressed as an integral over all plane wave directions. We include those directions for which the wavenumber is imaginary

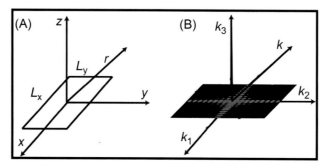

Figure 5.8 (A) Rectangular aperture in x, y, z cartesian coordinates. (B) Spatial transform of aperture in k-space. *From Szabo, T. L. (2014a). Overview. In Diagnostic ultrasound imaging: Inside out (2nd ed.). Elsevier.*

Table 5.1 Fourier transform acoustic variable pairs.

Variable	Transform variable	Type
Time t	Frequency f	$-i$
Space	Spatial frequency	
x	$\tilde{f}_1$	$+i$
y	$\tilde{f}_2$	$+i$
z	$\tilde{f}_3$	$+i$

(i.e., when $k^2 > k_1^2 + k_2^2$), where such wavenumbers correspond to nonpropagating waves that attenuate rapidly away from the source (see Eq. 5.4). The pressure is then written as the following integral

$$p(x, y, z) = p_0 \iiint_{-\infty}^{\infty} G(\hat{f}_1, \tilde{f}_2, 0)e^{i2\pi(\hat{f}_1 x + \hat{f}_2 y + \hat{f}_3 z)} d\tilde{f}_1 d\tilde{f}_2 d\tilde{f}_3 \qquad (5.5)$$

which because of Eq. (5.4), simplifies to

$$p(x, y, z) = p_0 \iint_{-\infty}^{\infty} G(\tilde{f}_1, \tilde{f}_2)\exp\left[i2\pi(\tilde{f}_1 x + \tilde{f}_2 y + \tilde{f}_3(\tilde{f}_1, \tilde{f}_2)z)\right] d\tilde{f}_1 d\tilde{f}_2. \qquad (5.6)$$

Notice that Eq. (5.6) is an integral over the plane defined by the $\tilde{f}_1(k_1)$ and $\tilde{f}_2(k_2)$ coordinates only, because $\tilde{f}_3$ is a dependent parameter as it can be expressed in terms of the other two components as in Eq. (5.4). Because the pressure p is obtained by summing over all plane wave directions, weighted by the function G, we call G the angular spectrum. In general, G may be complex and represent both the amplitude and phase of the plane waves.

For the specific example of a line aperture of length L along the x-axis, here is the corresponding spatial spectrum shown in Fig. 5.9, which turns out to be a sinc function from a transform like Eq. (5.2)

$$H(\tilde{f}_1) = \mathfrak{I}_{+i}[h(x)] = \int_{-\infty}^{\infty} \Pi(x/L_x)e^{i2\pi\tilde{f}_1 x} dx = L_x\text{sinc}(L_x\tilde{f}_1), \qquad (5.7)$$

from which the field pressure can be found,

$$p(x, z) = \int_{-\infty}^{\infty} A_x H(\tilde{f}_1)\exp\left[-i2\pi(\tilde{f}_1 x + \tilde{f}_3(\tilde{f}_1)z)\right] d\tilde{f}_1. \qquad (5.8)$$

Finally, Eq. (5.8) can be used to predict an acoustic beam in a two-dimensional plane.

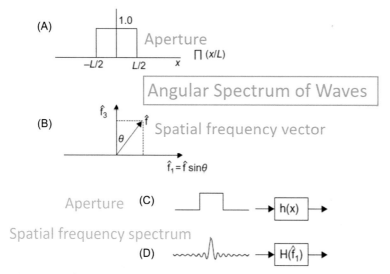

Figure 5.9 (A) Aperture function along *x*. (B) Spatial frequency vector. (C) Uniform amplitude aperture function. (D) Corresponding spatial frequency spectrum. *Adapted from Szabo, T. L. (2014a). Overview. In Diagnostic ultrasound imaging: Inside out (2nd ed.). Elsevier.*

5.2.2 Beamplot Simulator

The Beamplot Simulator is based on the angular spectrum equation, (5.8) with $\tilde{f}_1 = \tilde{f}\sin\theta$. A beamplot is a cross-section or slice of a beam along *x* at a distance *z*. This simulator calculates the pressure magnitude beamplot at any selected distance z: $|P[A(x_0), L_x, F, x, z]|$ as a function of wavelengths. The graphical panel is presented in Fig. 5.10 along with some explanatory comments about the GUI input parameters. Simplifications can be made since $kr = 2\pi(r/\lambda)$, so that all linear parameters are scaled by wavelength, such as $\hat{L} = L/\lambda$, etc. Input variables highlighted in Fig. 5.10 include an apodization function $A(x_0)$ (taper function over the aperture), aperture length, *L*, focal length, *F*, and field distance, *z*, all in units of wavelengths. The output is a symmetric beamplot and quantitative half-beam widths are also displayed in wavelengths.

In anticipation of the study of focusing effects, the Beamplot Simulator can demonstrate a surprising feature of focused fields, that in the focal plane (*x*, *y*, *F*) the beam cross-section is the Fourier transform of the aperture apodization or amplitude weighting function as illustrated in Fig. 5.11. A common motivation for the use of apodization is to reduce the level of the near-in sidelobes to make images clearer, as explained later.

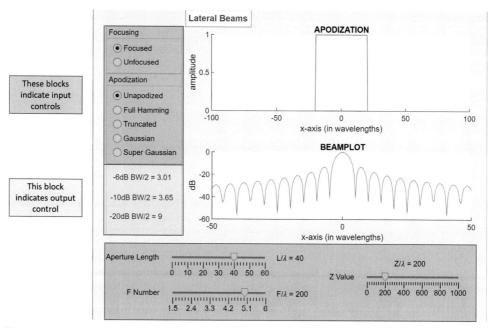

Figure 5.10 The Beamplot Simulator panel with inputs highlighted in red, and some output parameters highlighted in blue.

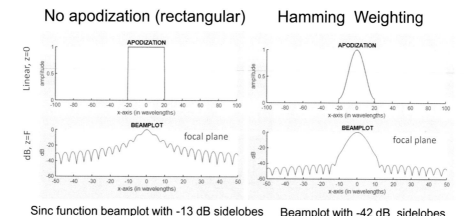

Figure 5.11 (Left) Rectangular apodization with focal plane beamplot below. (Right) Hamming weighting apodization with focal plane beamplot below.

5.2.3 Rayleigh integral model

Next we examine another useful equivalent model which can lead us to simplified ways of calculating and describing focused ultrasound beams. Instead of plane waves, this model uses spherical waves emanating from point sources. Each point in an aperture emits a spherical wave, so for example, if there are only two points, then the pressure is proportional to the sum of the waves from each point at $\mathbf{r}_1$ and $\mathbf{r}_2$,

$$p(\omega, \mathbf{r}) = \frac{\text{constant}}{2\pi} \times \left[\frac{e^{i[\omega t - \mathbf{k} \cdot (\mathbf{r} - \mathbf{r}_1)]}}{|\mathbf{r} - \mathbf{r}_1|} + \frac{e^{i[\omega t - \mathbf{k} \cdot (\mathbf{r} - \mathbf{r}_2)]}}{|\mathbf{r} - \mathbf{r}_2|} \right]. \tag{5.9}$$

Recall that sources of different geometries were originally introduced in Section 2.3.1 to create beams by the Expanding Waves Simulator. To extend the point source model shown in Fig. 5.5, an aperture can be thought of as containing a set of many infinitesimally tiny point sources radiating along the length or surface of an aperture. As the hemispherical wavefronts interfere constructively and destructively, more complicated wavefronts form and spill into regions beyond the aperture. At later times, or at farther distances, these individual wavefronts expand as illustrated using two points in Fig. 5.5.

In order to combine these wavefronts to produce the resulting field, a reference frame or set of coordinate systems is required. As shown in Fig. 5.12, a two-dimensional aperture is placed in a cartesian coordinate system with a vector $\mathbf{r}$ extending from the origin to a field point at (x, y, z). In order to sum the infinitesimal source fields throughout the aperture at points $(x_0, y_0, 0)$, consider a vector, $\mathbf{r}_0$, from a source point to the field point position,

$$p(\mathbf{r}, \omega) = \frac{i\rho_0 c k v_0}{2\pi} \int_s \left\{ \frac{e^{i[\omega t - \mathbf{k} \cdot (\mathbf{r} - \mathbf{r}_0)]}}{|\mathbf{r} - \mathbf{r}_0|} \right\} A(\mathbf{r}_0) \, dS, \tag{5.10}$$

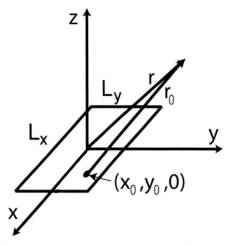

Figure 5.12 Field coordinates for a rectangular aperture. *From Szabo, T. L. (2014b). Beamforming. In Diagnostic ultrasound imaging: Inside out (2nd ed.). Elsevier.*

where $v_n = v_0 A(\mathbf{r}_0)$ is the normal particle velocity and $A(\mathbf{r}_0)$ is its distribution across the aperture S. The term in braces can be recognized as the field from a single point source like those in Eq. (5.8). This equation allows the calculation of the diffracted pressure field in the half-space $z > 0$ above the aperture. Eq. (5.10) states that the pressure at point $\mathbf{r}(x,y,z)$ is the sum (integral) of the fields of point sources distributed finely over the aperture, with a complex amplitude weighting function $A(\mathbf{r}_0)$ for each point source at point r_0. A hemispherical wavefront originating from a point source at $\mathbf{r}_0$ is described by the parenthetical term in the integrand. This equation is known as the Rayleigh integral, which expresses the spherical wavefront principle in mathematical form. Because the pressure is expressed at a single frequency $\omega = 2\pi f$, this is the continuous waveform or frequency domain form of the Rayleigh integral.

Sometimes the aperture function is "separable," that is, this two-dimensional function can be written as the product of two one-dimensional functions. Then the field expressions for the xz plane can be written without any y-dependence. This simplification results from separating the aperture function $A(\mathbf{r}_0)$ into the product of two functions, each a function of a single variable, such as the rect functions illustrated in Fig. 5.13,

$$A(\mathbf{r}_0) = A_x(x_0)A_y(y_0). \tag{5.11}$$

To see the advantage of this approach, refer to a typical one-dimensional ultrasound array which is the rectangular aperture in Fig. 5.13 with the zy plane rotated 90 degrees clockwise so that the z-axis is pointed to the right as in Fig. 5.14. The extent of the array along the x-axis forms the L_x aperture which radiates into the positive z-axis half-space, but we will only examine the field in the cut plane xz, also known as the "azimuth plane." Along the y-axis is the L_y aperture which leads to a simple expression for the field in the yz plane, called the "elevation plane." The beams in both orthogonal planes are focused by

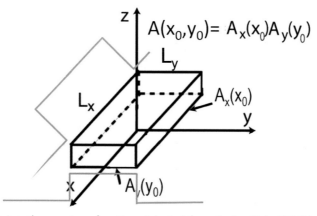

Figure 5.13 Factoring the aperture function. *Adapted from Szabo, T. L. (2014b). Beamforming. In Diagnostic ultrasound imaging: Inside out (2nd ed.). Elsevier.*

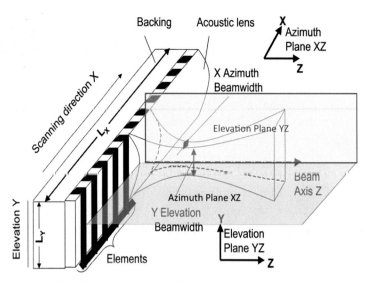

Figure 5.14 Imaging field: azimuth or imaging plane with *xz* field created by electronic focusing and *yz* or elevation plane where focusing is typical with a mechanical lens (Panda, 1998). *Adapted from Panda, R. K. (1998).* Development of novel piezoelectric composites by solid freeform fabrication techniques *(Dissertation). New Brunswick, NJ: Rutgers University.*

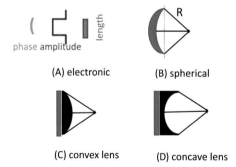

Figure 5.15 Types of focusing: (A) Electronic options. (B) Spherical. (C) Convex lens. (D) Concave lens. The blue represents the active aperture and black represents a lens material.

different mechanisms: electronically by the array in the *xz* plane and by a mechanical lens in the *yz* plane. The 3D field can be expressed as the product of the 2D fields computed from separable portions of the aperture function.

To examine the options available in an array to control the beam characteristics, assume the aperture along *x* is creating a beam that propagates along *z* and its diffracted beam fills the two-dimensional plane *xz*. For now, consider the array to be represented by a solid aperture (of the same overall size) which has a length, amplitude, and phase, as shown in Fig. 5.15A. The aperture is illuminated with an amplitude shape $A(x_0)$ and length L_x. Here the shape shown is rectangular ($\Pi(x/L_x)$, constant amplitude over a length L_x) as shown in Fig. 5.11. It can be other shapes such as Gaussian or half-period cosine. Changing amplitude shape is called "apodization" as already demonstrated by the Beamplot Simulator (Figs. 5.10 and 5.11).

A way of deliberately reducing beamwidth is called "focusing." Most often focusing is implemented by producing a curved wavefront that converges at a point F along the acoustic beam axis. For example, in Fig. 5.15B, the path lengths from points along the curved inner surface of the transducer to a focal point F are the same since they arrive at the center of the radius of curvature of the transducer. In general, from a flat transducer, the travel times to the acoustic axis are longer from the ends of the aperture than from the center.

Therefore by increasing the phase delay across the aperture (maximum in the middle), the phased path lengths along the aperture are adjusted to arrive coherently at point F. For the continuous wave case, which we are now considering, phase change is implemented by changing the relative phase across the aperture (i.e., moving the waves forward or backward). In Chapter 7, focusing is described in terms of adjusting the relative delays of pulses arriving at individual array elements.

Focusing is accomplished by altering the phase across the aperture. For an array, the aperture length, apodization, and focusing are accomplished electronically as in Fig. 5.15A. Mechanical focusing can take the form of shaping the aperture such as the curved transducer in Fig. 5.15B or using the lenses in Fig. 5.15C and D. In the simplest case, a spherically curved transducer, as in Fig. 5.15B, a phase change is added geometrically so the source is

$$H(r_0) = A(r_0)\exp(ir_0^2/2R), \tag{5.12}$$

where $r_0^2 = x_0^2 + y_0^2$. In one dimension, a general expression for the source required for focusing is approximately

$$H(x_0) = A_x(x_0)\exp(ix_0^2/2F), \tag{5.13}$$

where F is the focal length. In one dimension for a cylindrical lens like the one in Fig. 5.15B, $F = R$, the radius of curvature. This is the same type of phase curve that would be produced for electronic focusing in Fig. 5.15A. For the two mechanical lenses,

$$F = \frac{R_{lens}}{n-1}, \tag{5.14}$$

where n is the index of refraction, $n = c_W/c_L$, where c_L is the speed of sound in the lens and c_W is the speed of sound in the surrounding water or tissue, and R_{lens} is the radius of curvature of the lens. Unlike optics, acoustics can utilize lens materials with a sound speed that can be either greater or smaller than the surrounding medium, and therefore it can use both convex and concave shapes to create a convergent, focused wavefield. In the typical optical case, the speed of light in the lens is less than that in the surrounding medium (air), so the index of refraction is greater than 1. In acoustics if $n > 1$, the lens is a convex shape as in Fig. 5.15C; but if $n < 1$, the lens is concave, as in Fig. 5.15D.

An explanation for focusing in terms of waves is more involved. A ray model such as that for the spherical transducer in Fig. 5.15B is straightforward: the rays from different parts of the aperture all converge at the intense focal point. Without focusing, apertures on the order of tens of wavelengths wide diffract into a predictable pattern such as that in Fig. 5.3. This outward unraveling and interference of spherical wavefronts like those in Fig. 5.5 can be described by a positive phase advance of the wavefront in terms of a combined positive quadratic and linearly advancing negative phase term with distance. Because of the convention used for a positive going wave outside the integral, the signs of the actual phase terms are opposite; for example, a negative quadratic term corresponds to a expanding positive wavefront. Focusing is an attempt, by design, to introduce a negative *physical* quadratic phase, Eq. (5.13), to cancel the quadratic phase from naturally occurring diffraction at the focal length. At this one distance, $z = F$, the quadratic phases cancel, and only linear phase is left so that the beamshape at the focal plane is the spatial Fourier transform of the aperture function (the Fresnel approximation of the Rayleigh integral, Eq. 5.10), also available in (Szabo, 2014b).

$$p(x, y, z) = \frac{ip_0 k}{2\pi z} e^{i(\omega t - kz)} e^{-ik(x^2 + y^2)/2z} \int\int_s \left[e^{-ik(x_0^2 + y_0^2)/2z} e^{ik(x_0^2 + y_0^2)/2F} A(x_0 y_0) \right] e^{ik(xx_0 + yy_0)/z} dx_0 dy_0.$$

$$(5.15)$$

Thus, neglecting plane wave terms outside the integral,

$$p(x, z) = \sqrt{\frac{ip_0 k}{2\pi z}} \int \left[e^{-ikx_0^2/2z} e^{ikx_0^2/2F} A_x(x_0) \right] e^{i(kx/z)x_0} dx_0;$$

$$(5.16)$$

the first two terms in the integrand cancel at $z = F$, leaving the spatial scaled Fourier transform,

$$p(x, F) = \sqrt{\frac{ip_0 k}{2\pi F}} \int A_x(x_0) e^{i2\pi(x/\lambda F)x_0} dx_0.$$

$$(5.17)$$

For an unapodized aperture, $A(x_0) = \Pi(x_0/L_x)$ so this $+ i$ spatial Fourier transform with variable x_0 in one domain is transformed to a function of variable $(x/\lambda z)$ in the other domain,

$$p(x, z) = \sqrt{\frac{ip_0 k}{2\pi z}} L_x \text{sinc}\left[L_x\left(\frac{x}{\lambda z}\right)\right]. \tag{5.18}$$

Note this equation can be interpreted in several ways. The beamplot is plotted against the wavelength-scaled variable, $\hat{x} = x/\lambda$, as was done in Fig. 5.11 or it can be thought of in terms of a wavelength-scaled aperture, $\hat{L}_x$. For a beamplot at the focal length distance, substitute $z = F$ in Eq. (5.18). Much more information on these topics can be found in Szabo (2014b). For the focal gain on the beam axis, $x = 0$, and the sinc function equals one. Finally, the pressure in a focal plane can also be obtained for $z = F_x = F_y$ from Eq. (5.15)

$$p(x, z) = \frac{ip_0 k}{2\pi z} L_x \text{sinc}\left[L_x\left(\frac{x}{\lambda z}\right)\right] L_y \text{sinc}\left[L_y\left(\frac{y}{\lambda z}\right)\right], \tag{5.19}$$

where, as in Eqs. (5.17) and (5.18), the variable z is retained to apply to either the focus, $z = F$, or z in the far field.

An unfocused field such as that depicted in both Figs. 5.3 and 5.4 has a "near field" and a "far field." The change occurs at a transition distance in a plane at approximately

$$z_{\max} = \frac{L_x^2}{\pi\lambda}. \tag{5.20}$$

For the 40 wavelength aperture for the field in Figs. 5.3 and 5.4, this transition occurs at 509λ. As evident in Fig. 5.3, this is the distance in which the beam transitions into a more far-field sinc shape for this unapodized case. Note from Eqs. (5.18) and (5.19) it is evident that in the far field, the beam continues to expand in width and fall in amplitude as z increases. For a focused beam, there is no near or far field. Instead there is a focal region where the beam is relatively narrow preceded by a prefocal region and afterward, a postfocal region as explained in more detail in (Szabo, 2014b).

5.3 Field Simulator

5.3.1 Field Simulator control panel

Because the Field Simulator is the most complicated of the simulators discussed so far, it is worth reviewing variables and settings. Several input variable controls combine to

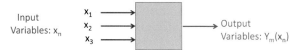

Figure 5.16 Simulator variables: inputs transformed to outputs.

create the output variables, which in this case is a pressure or intensity field in a plane, as graphically represented in Fig. 5.16.

To summarize the Field Simulator so far, we have the following 2D wavelength-scaled input variables: $A(\hat{x}_0, \hat{L}_x, \hat{F})$. A represents a selectable apodization function which by default is rectangular. The output is pressure amplitude (or intensity which is proportional to pressure squared) depending on the selected equation or model. The pressure is specified at a point(x,z); therefore output pressure is a function of input variables: $P(\hat{x}_0, \hat{L}_x, \hat{F}, \hat{x}, \hat{z})$.

Unlike the Beamplot simulator which used the Angular Spectrum of Waves model, the Field Simulator employs the continuous wave (single temporal frequency) Rayleigh Integral of Eq. 5.16. The two models give nearly equal results, where the differences arise from numerical approximations; physically, the models are equally valid decompositions of a field. The use of a particular model is transparent in a simulator—detailed knowledge, or mathematical manipulation of a model is not needed to use the simulator.

The input variable controls are identified by orange text in Fig. 5.17. Output display options provide ways of visualizing the resulting calculations of the model: for example, Linear Field (or Stacked) produces individual beamplots (cross-sections of beam amplitude) separated by a distance interval, $\Delta\hat{z}$, that are arranged above each other with offsets to enhance their visualization, as done in Fig. 5.17. Output display options of the Field Simulator graphical user interface are described by text in green. Note variables x and z do not have to be explicitly stated as they are automatically determined to provide plots of the interesting parts of the field.

5.3.2 Beamplot focusing characteristics

The Field Simulator not only offers different display options such as the stacked beam plots in Fig. 5.17 and the surface plot in Fig. 5.4, but also provides some interactive options. An example of an interactive option is shown in Fig. 5.18 in which an intensity plot is shown. The interactive option is enabled by clicking and dragging a horizontal line through the beam so that a beam plot is created as shown in Fig. 5.19. Similarly, a vertical line can produce an axial plot. There are simple relations for

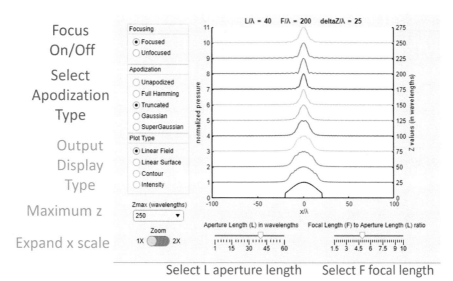

Figure 5.17 Field Simulator graphical user interface with annotated input controls and selections and display options.

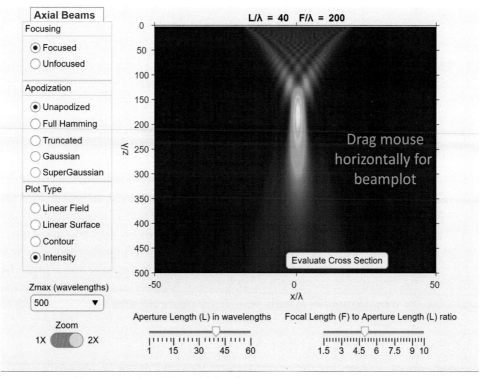

Figure 5.18 Field intensity plot of the Field Simulator.

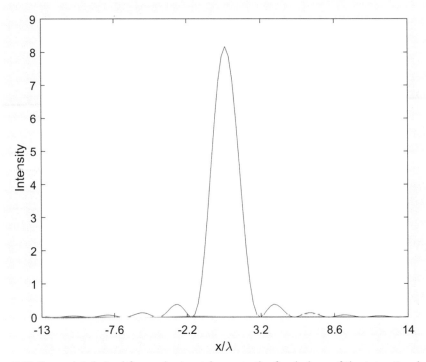

Figure 5.19 Beamplot derived from a horizontal scan in the focal plane of the intensity plot in the field simulation from Fig. 5.18.

fullwidth half-maximum widths for an unapodized aperture, derived from Eq. (5.18), and other equations are possible for different apodizations,

$$\text{FWHM} = 1.206\lambda z/L. \tag{5.21}$$

Of interest is that the beam shape in the focal plane is the Fourier transform of the aperture function as illustrated in Fig. 5.19. The beamwidth here is determined by the half-amplitude points of a sinc function, Eq. (5.18). This shape is recognizable as the Fourier transform of a rect apodization function. $\hat{z} = \hat{L}^2/\pi$ In the far field, the shape is also the Fourier transform of the aperture function, so Eq. (5.18) works there also. It is useful to know that beams have the same shape if they have the same value of S, the universal scaling parameter, where $S = \lambda z/L^2$. In Eq. 5.18 we can show that the field shape (thus the beamplot) scales linearly with L_x by substituting $\lambda z = SL_x$.

5.3.3 Axial focusing characteristics

The axial intensity indicates the strength of focusing if present. It can be obtained by dragging a mouse vertically along the beam axis instead of horizontally as in Fig. 5.18, so that an axial intensity plot results as shown in Fig. 5.20. Focusing, examined here in the xz plane identified in Fig. 5.14, simultaneously narrows the beam and causes a peak on the axis.

Focusing can be quantified by several parameters. The F-number, $F\#$, for apertures at $z = 0$ in the xz plane is

$$F_x\# = F_x/L_x,\tag{5.22}$$

and there is a corresponding $F_y\#$ number for the yz plane. The smaller the $F\#$ is, the stronger the focus.

For a rectangular aperture, two different focusing mechanisms may be involved in the orthogonal planes xz and yz, so, initially, a circularly symmetric aperture with a radius a is easier to understand. This kind of unapodized aperture has an area equal to πa^2. The peaks produced by this geometry at the focal length F have a focal gain,

$$G_{\text{focal}} = \pi a^2/\lambda F = \text{area}/\lambda F,\tag{5.23}$$

which is the ratio of pressure at F over that on the aperture: $p(F)/p(0)$. Plots of axial pressures for three different focal lengths are given in Fig. 5.21. Note that the peaks

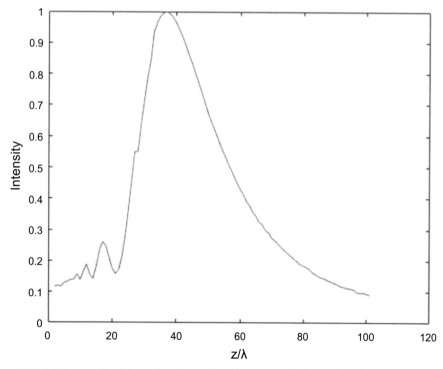

Figure 5.20 Axial normalized intensity plot derived from a vertical scan line along the z-axis in the intensity plot from the Field Simulator plot in Fig. 5.18.

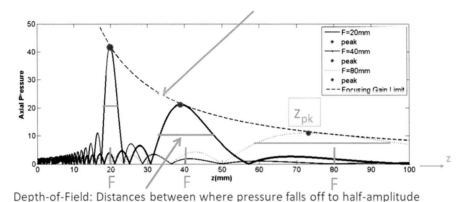

Peak pressures occur at z<F or z_{pk}/F decreases as F increases

Peak pressure at the focal length F falls off as 1/F for the same Aperture

Depth-of-Field: Distances between where pressure falls off to half-amplitude increases as F increases

Figure 5.21 Axial pressures for three different focal lengths and a spherical aperture of $a = 10$ mm. *From Szabo, T. L. (2014b). Beamforming. In* Diagnostic ultrasound imaging: Inside out *(2nd ed.).* Elsevier.

fall under Eq. (5.23) as an upper bound and so that the shorter the focal length, the higher the peak. As the peak value increases, its extent or depth-of-field (distance between half-amplitude axial points) decreases. As F increases, note that the location of the peak is less than F, as explained in Szabo (2014b). Also, a focused beam can be thought of having three regions: a focal region where the beam remains narrow over what is called the "depth of field," and before it, a prefocal region, and past it, a postfocal region (Szabo, 2014b). For rectangular apertures, there are two different focal mechanisms and lengths in general. Only when they are coincident does Eq. (5.18) hold as the product of the focal gain in each plane (from a version of Eq. (5.19) at $x = y = 0$, $z = F_x = F_y$),

$$G_{\text{focal}} = L_x L_y / \lambda \sqrt{F_x F_y} = \left(L_x / \sqrt{\lambda F_x}\right)\left(L_y / \sqrt{\lambda F_y}\right). \qquad (5.24)$$

To distinguish the differences in focusing properties in different planes such as xz and yz, because L_x and L_y are different, an alternate definition of depth of focusing is based on beamwidths. This depth is defined as the difference between two axial distances, where the -6 dB beamwidths increase over the minimum beamwidth by a factor of 2 (IEC, 2020; Szabo, 2014b) in a plane.

A two-dimensional view of the extent of a focal region is illustrated by the Field Simulator in Fig. 5.22. Here the beam is displayed as a contour plot for an approximately 40 wavelength long aperture and an $F_x\# = 5$. The inner -6 dB contour provides an image of this region in both its axial and lateral extent. The contours are referenced to the maximum field value at each depth, whereas the intensity plot is referenced to the global maximum in the plot.

In general, when apodization is used, the focal gain at a coincident focal point, $F_x = F_y$ can be derived from Eq. (5.15),

$$G_{\text{focal}} = \left| p(0, 0, F)/p_0 \right| = \frac{1}{\lambda F} \int \int_s [A(x_0, y_0)] dx_0 dy_0, \tag{5.25}$$

which for separable apodization schemes, A_x and A_y reduces to the known focal gain for unapodized apertures, Eq. 5.24.

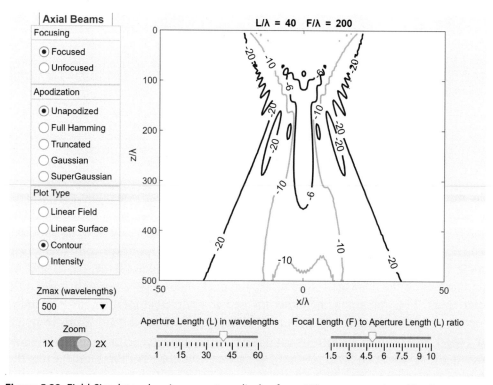

Figure 5.22 Field Simulator showing a contour display for a 40λ aperture and an $F\# = 5$.

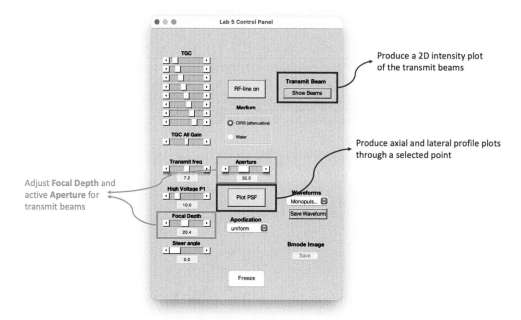

Figure 5.23 Lab 5 Vantage™ GUI, adding the transmit beam plotting tool ("Show Beams" button) and the beam profiling tool, with the button designated "Plot PSF." Also new for this lab are the Focal Depth and Aperture sliders to control transmit beam shape.

5.4 Lab 5: beams and focusing and the point spread function

5.4.1 Exercises with the physics simulators

The first of the simulators is the **Beamplot Simulator** which computes a continuous wave lateral, or transverse, field profile at an adjustable distance from the source. The source field is a one-dimensional aperture of adjustable length and may be apodized using several classical windows or weighting functions. This simulator computes a field profile like one might measure in a tank by scanning a hydrophone, or, as will be illustrated using the Vantage™ system, a transverse section taken from an image of a point target. The source may be either focused at a selectable focal length F or unfocused (simulated by setting a very long focal length). Students are guided to compute the lateral field profiles for various combinations of parameter settings and to observe directly the Fourier transform relationship between the shape of the source across the aperture (apodization) and the field at the focus. They are also instructed to study the evolution of the field structure, using this one-dimensional view, from the prefocal field through the focal region and then on to the postfocal field for the different weighting functions and F-number ($F\#$) conditions. They will observe the tradeoff

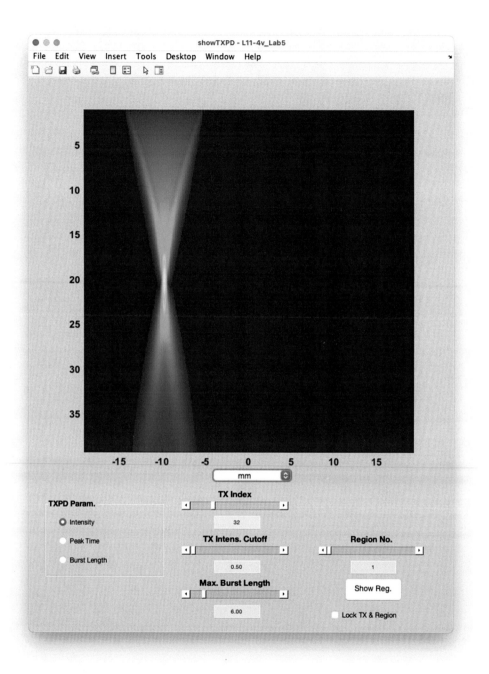

Figure 5.24 Vantage™ transmit beam intensity visualization tool. Here, one of 128 focused beams used to construct a single image is displayed.

between sidelobe strength and focal width as a function of apodization, and will quantify the increase in focal spot size with increasing $F\#$.

The **Field Simulator** uses the same type of continuous wave source but computes the field over the two-dimensional plane, xz. Several different display options reinforce the understanding of the field structure using different geometrical approaches, including a waterfall plot of lateral profiles like those observed earlier but one at a time, a 3D surface with field intensity represented by height and color, an intensity color map in the xz plane, and a contour plot that facilitates quantitative interpretation.

Through the exercises, students learn about simple relations that estimate the focal width given aperture and focal distance. They are introduced to the concept of depth of field and can quantify it using the contour plot. They observe that within a range of F-numbers, the field structure for one $F\#$ can be obtained by a linear stretching transformation of the field for another $F\#$. They can obtain axial profiles as well and observe prefocal field and postfocal zones on either side of the focal region.

5.4.2 Experiments and exercises with the Vantage™ system

The graphical interface for the Vantage™ now features additional controls to allow the student to adjust focal depth and aperture width and select an apodization function for the beams, such as the one in Fig. 5.24, used to create the B-mode image in Fig. 5.25.

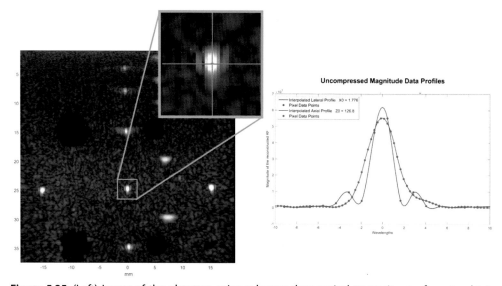

Figure 5.25 (Left) Image of the phantom using a longer than typical transmit waveform to obtain a quasicontinuous wave result, with a zoomed-in portion in which the intensity profiles are collected. (Right) Plot of the axial and lateral intensity profiles, interpolated from the underlying unprocessed image data.

These beams can be visualized in a way that is now familiar to the students by using the transmit beamplot Map simulation tool that is built into the Vantage™ software, as shown in Fig. 5.24. This tool is not intended to reproduce all aspects of the continuous wave field because it is better suited to pulsed beams, the subject of Chapter 7, and in more detail in (Szabo, 2014c).

In addition, a button is provided to collect a "field map" in a small region of the image by selecting a target such as one of the point targets in the imaging phantom. The program launched by the "Plot PSF" button finds the brightest point near the selected image location and displays a crosshair over that point. The image data for the axial and lateral axes represented by the crosshairs is extracted not from the image pixel themselves, but from the underlying data prior to signal processing that uses logarithmic compression and 256 gray levels on the display window; depending on image brightness and contrast these display pixel values may be saturated or significantly processed to improve image appearance. To best compare with the simulator results, the unprocessed profile data is displayed on a graph, producing an axial pulse envelope and a lateral beamplot as shown in Fig. 5.25. The student can then use MATLAB®'s built-in data display tool to select any point on the plots to obtain quantitative measures at that image location. Repeating this procedure for different locations in the image (e.g., prefocal, near focal, and postfocal), and for different beam settings, the concepts explored using the simulators can be validated experimentally. Students are encouraged to go back and forth between simulation and experiment.

References

IEC 61828. (2020). *Ultrasonics − focusing transducers − definitions and measurement methods for the transmitted fields* (Ed. 2.0). International Electrotechnical Commission.

Panda, R. K. (1998). *Development of novel piezoelectric composites by solid freeform fabrication techniques* (Dissertation). New Brunswick, NJ: Rutgers University.

Szabo, T. L. (2014a). Overview. in *Diagnostic ultrasound imaging: Inside out* (2nd ed.). Elsevier, Chapter 2.

Szabo, T. L. (2014b). Beamforming. in *Diagnostic ultrasound imaging: Inside out* (2nd ed.). Elsevier, Chapter 6.

Szabo, T. L. (2014c). Array beamforming. In *Diagnostic ultrasound imaging: Inside out* (2nd ed.). Elsevier. Chapter 7.

CHAPTER 6

Continuous wave array beamforming and heating

6.1 Overview

6.1.1 Beamformers in the block diagram

The blocks in Fig. 6.1 covered in this chapter still include diffraction but add the transmit and receive beamformers to control the beam. The apertures themselves are now arrays with individually controllable elements. Finally, a path for absorption is added by which more realistic propagation in tissue and other materials can be modeled.

6.1.2 Array beamforming

Previously in Chapter 5, we encountered solid apertures, tens of wavelengths long, which radiated acoustic beams. Diffraction was explained as the interference pattern of many tiny spherical wavefronts emitted by a solid aperture. Similarly, finite-size point sources can be arranged in regular geometries to form arrays which in turn can form differently shaped beams from the interference of their individual point source wavefronts. Examples of these arrangements can be found in Fig. 6.2. A flat two-dimensional (1D) array can generate a finite-sized approximation of a plane wave. A 1D curved array can focus on a point in-plane. Depending on the phasing of the points, a line of points can radiate circumferentially. In another configuration, points can be phased to direct or steer a beam in a vector direction. Finally, the elemental point source individually radiates an expanding spherical wave. These are simplifications, of course, because the wavefronts generated by these points will change with distance. In this chapter, we will study the beams formed by arrays of sources of finite extent.

6.1.3 Wavefronts Simulator

The Wavefronts Simulator is based on a simple equation: the sum of spherical waves,

$$p(\mathbf{r}) = \sum_n \frac{\exp i[\omega t - k(\mathbf{r} - \mathbf{r}_n) - \theta_n]}{4\pi(\mathbf{r} - \mathbf{r}_n)} \tag{6.1}$$

where for pressure at vector position $\mathbf{r}$ the spherical waves from source positions $\mathbf{r_n}$ which may have a phase θ_n applied to each one.

Essentials of Ultrasound Imaging
DOI: https://doi.org/10.1016/B978-0-323-95371-9.00004-7

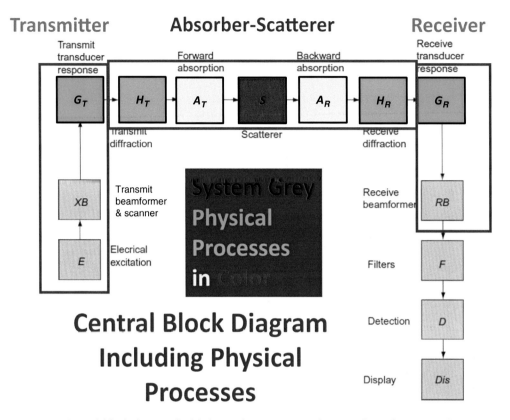

Figure 6.1 Central block diagram highlighting the transmit and receive beamformers and transmit and receive diffraction, and absorption paths including a point scatterer (Szabo, 2014a). *Adapted from Szabo, T. L. (2014a). Overview. In* Diagnostic Ultrasound Imaging: Inside Out *(2nd ed.). Elsevier.*

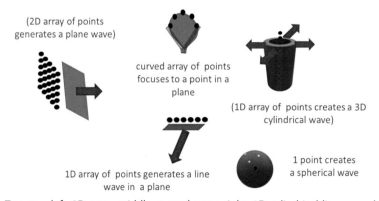

Figure 6.2 Top row: left: 2D array, middle: curved array, right: 1D cylindrical line array; bottom row: left 1D line array, right: point source.

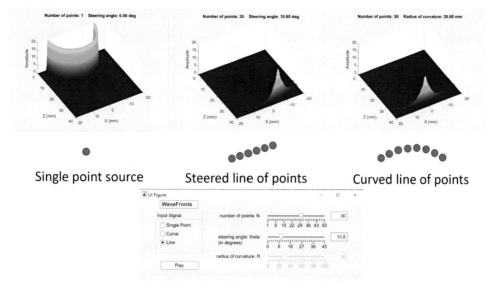

Figure 6.3 Snapshots from the Wavefronts Simulator: Top: (left) a single source, middle, a line of point sources with a steering option shown and right, a set of point sources arranged on a selectable radius of curvature. Below, the Wavefronts Simulator graphical user interface panel.

The Wavefronts Simulator of Fig. 6.3 provides an option to select a single point, n points in a straight line, or a curve of n points spaced half a wavelength apart. In addition, for a curve, the radius of curvature can be chosen. Adding a steering angle is another option.

The Wavefronts Simulator GUI panel is shown in the lower half of Fig. 6.3. The three options are illustrated above: on the left is a single point source, next is a line source with 30 points with linearly delayed onset times to produce a steered beam, and finally in the upper right, is a curved arrangement of sources with a selectable radius of curvature (a smaller subset of points is drawn for clarity).

With the aid of the Wavefronts Simulator, we can visualize how beams are formed. This simulator launches simultaneously a set of expanding spherical wavefronts from the sources; each wavefront is represented by an impulse-like pulse in time for easier visualization. Animations from a line source with only four point sources show a remarkable result in Fig. 6.4. With so few sources, the spherical wavefronts coalesce into a recognizable beam very near the sources. Farther from the sources, the spherical overlap increases and the beam widens.

6.1.4 In this chapter you will learn

Arrays are the key to controlling acoustic beams in real time. Arrays have individually addressable elements arranged in a regular pattern. Apodization and focusing can be

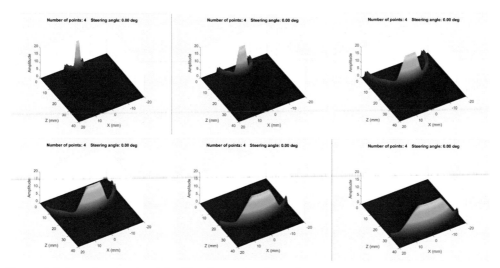

Figure 6.4 An animation from the Wavefronts Simulator demonstrating the composite wavefront developing in a time sequence from a line of four equally spaced point sources.

implemented in an array by changing the amplitude and phase of a signal sent to each element. Because elements are commonly realized in the shape of three-dimensional (3D) cuboids, they are imperfect samplers, imprinting artifacts and limitations on array performance. The basic principles of an array can be explored in the form of arrays of point sources modeled by the Wavefronts Simulator. Each element in the array affects the overall directivity through an element factor. The important consequences of adjusting the size of an element and its spacing on array directivity can be investigated by the Directivity Simulator. A more realistic exploration can be undertaken with the CW Array 3D Simulator which calculates the full 3D beam radiated by a 1D array of cuboid elements (3D in shape). Additional display options in the simulator allow examination of the acoustic field away from the principal planes. The option of choosing lossy media provides an opportunity for in-depth study of how absorption in materials alters the beamshape. A byproduct of absorption is heating, which can also be calculated by the simulator and compared to the acoustic field. Finally, an alternative method of focusing called "plane wave compounding" can create a pseudo-focusing effect by combining pairs of symmetrically steered pairs of beams. This approach can be examined through the Plane Wave Compounding Simulator.

6.2 Imperfect element samplers

6.2.1 Rectangular array elements

In order to characterize the operation of a real phased array for ultrasound imaging, we begin with ideal samplers. Point sources (or receivers) can be considered to be perfect spatial samplers. Now consider a plane wave arriving at an angle θ from the z-axis at an array of point sources spaced at distances p from each other as in Fig. 6.5. The difference in path length from element to element is $p\sin\theta$. Therefore the path difference between three elements is $2p\sin\theta$ and so on. A useful variable is $u = \sin\theta$.

The simple finite line source of n point sources depicted in Fig. 6.3 can be considered to be an infinite series of point sources multiplied by a rectangular function just wide enough to capture n point sources. The translation of this concept into mathematical terms is illustrated in Fig. 6.6. On the left side is the multiplication of a rectangle function multiplied by an endless sequence of impulse functions representing the point sources. The series of impulses is transformed into another series in the other domain with a different periodicity. The spatial Fourier transform of a rectangle function is a sinc function. So, in this figure on the left in the space domain along x the multiplication by a rectangle becomes, in the other domain, a convolution of a sinc function with an infinite series of impulses which is seen on the right side of the figure as

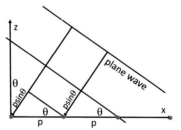

Figure 6.5 A plane wave arrives at an angle θ to an equi-spaced array of point samplers. The path difference from element to element is $p\sin\theta$.

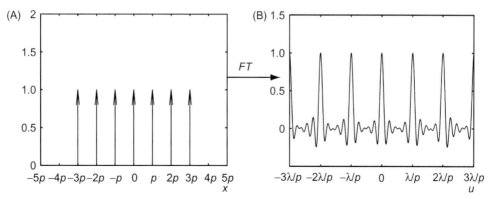

Figure 6.6 (A) A finite number of point sources spaced at intervals p form a sampled aperture which transforms into (B) an unending series of sinc functions as a function of u (Szabo, 2014b). *From Szabo, T. L. (2014b). Array beamforming. In* Diagnostic Ultrasound Imaging: Inside Out *(2nd ed.).* Elsevier.

endlessly repeated sinc functions. For this example in Fig. 6.6, the rectangular aperture of length L_x is sampled seven times at a periodicity p, which, when spatially transformed, becomes

$$\Im_i\left[\prod\left(\frac{x}{L_x}\right)\mathrm{III}\left(\frac{x}{p}\right)\right] = pL_x\sum_{-\infty}^{\infty}\mathrm{sinc}\left[\frac{L_x}{\lambda}\left(u - m\lambda/p\right)\right] \tag{6.2}$$

in which $u = \sin\theta$, $\lambda = c/f$, and **III** is the Shah function (Bracewell, 2014) (an unending series of impulse functions, whose argument is scaled by the pitch p, the spacing between impulses, and whose spatial Fourier transform is another Shah function in the spatial frequency domain). In this case, the Shah function is effectively multiplied by a finite aperture rectangular window of width L_x. By the convolution theorem, we can see that the Fourier transform of the windowed impulses is given by the convolution of the transforms of the rectangular window (the sinc function) and an infinite series of impulses in the frequency domain spaced by λ/p. From this relation, the width of each of the repeating sinc functions is determined by the aperture in wavelengths. We will return to the complications caused by this sampling operation later.

Most often an ultrasound array is not made of ideal point samplers arranged in a row but rectangular elements with shapes called "cuboids" by mathematicians. As shown in Fig. 6.7, the dimensions of an element along x, y, and z are width, w, length, L_y, and depth, d. Because the elements in the array are placed along x with a period or pitch, p, this is called a "1D" array even though it exists in a 3D physical form as discussed in Section 5.2.3. The lower right of Fig. 6.7 shows the graphical representation of this 1D characteristic along x.

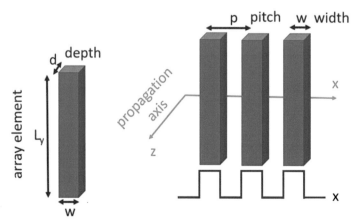

Figure 6.7 The size and arrangement of cuboid elements of a one-dimensional array and their representation along x.

6.2.2 Sampling by elements

The influence of the shape of an element on the sampling process is illustrated in Fig. 6.8. As indicated in the bottom right of Fig. 6.7, the sampler assumes a rectangular shape along x. Each element is a tiny solid aperture along x with typically less than a half-wavelength or at the most two wavelengths in width w. This small size, $L_e = w$, creates a far-field pattern within a short distance away; therefore the 1D spatial Fourier transform of this element shape is the familiar sinc function as shown in Fig. 6.8. The element in the long direction, L_y, however, can be tens of wavelengths in length, so its far-field spatial transform, also a sinc function, is much narrower in angle space. In this case, the far field in the yz plane occurs much farther away than for an individual element in the xz plane. We will later see that a focusing lens is usually applied in this yz plane to mechanically focus the beam and thus produce a thin imaging region (see Fig. 6.11, Y-elevation beamwidth).

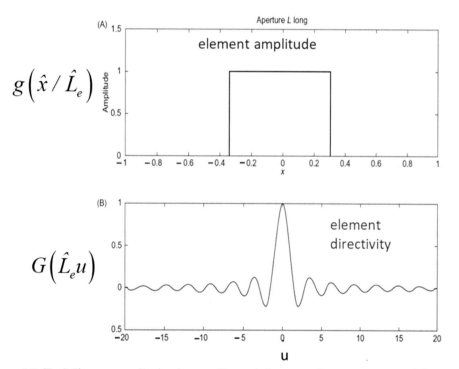

Figure 6.8 (Top) Element amplitude along x. (Bottom) Corresponding aperture spatial frequency spectrum (Szabo, 2014a). The hatted variables are wavelength normalized. *Adapted from Szabo, T. L. (2014a). Overview. In Diagnostic Ultrasound Imaging: Inside Out (2nd ed.). Elsevier.*

6.3 Array directivity

6.3.1 Array and element factors

To generate the overall array pattern shown in the bottom right of Fig. 6.7 one simply convolves the impulse samplers with the element amplitude function g of Fig. 6.8. Applying the convolution theorem results in the angular spectrum displayed in Fig. 6.9 and demonstrates that the overall array directivity is the product of the repeating sequence of sinc functions from the right side of Fig. 6.6, the "array factor," modulated by the much wider element directivity or "element factor." The value of $u = 0$ corresponds to the direction of the z-axis; whereas $u = \pm 1$ corresponds to the directions at $\pm y$ axes or ± 90 degrees. Shown in Fig. 6.9 is the case for $p = \lambda$ and $w = 0.5$ and $n = 7$; however, for illustrative purposes, only three elements are shown. This simplification is to illustrate the relationship among variables from Fig. 6.9A: the

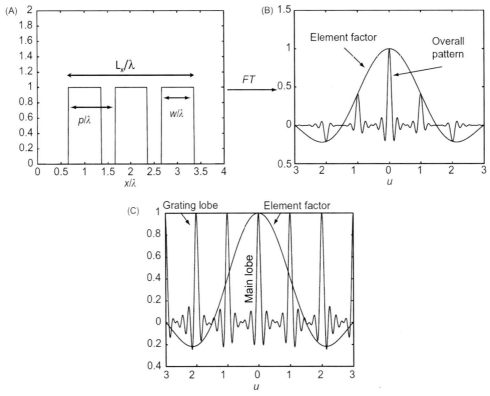

Figure 6.9 (A) Array made of elements w wide and spaced p apart, (B) overall array directivity pattern showing modulation by element factor, and (C) individual array factor with grating lobes and element factor (Szabo, 2014b). *From Szabo, T. L. (2014b). Array beamforming. In* Diagnostic Ultrasound Imaging: Inside Out *(2nd ed.). Elsevier.*

periodicity p is associated in the other domain, Fig. 6.9B, with the locations of the grating lobes, w is associated with the element factor and L_x is linked to the widths of the sinc grating lobe patterns.

The consequences of sampling here are the extra artifactual grating lobes at $u = \pm 1$. The location of the grating lobes can be predicted by

$$\theta_g = \pm \arcsin\left(m\lambda/p\right) \tag{6.3}$$

This result is a consequence of many values for the integer m in the overall directivity equation,

$$\Im_i\left[\left\{\Pi\left(\frac{x}{w}\right)\right\} * \Pi\left(\frac{x}{L_x}\right)\mathrm{III}\left(\frac{x}{p}\right)\right] = \left\{w\mathrm{sinc}\left(\frac{wu}{\lambda}\right)\right\}pL_x\sum_{-\infty}^{\infty}\mathrm{sinc}\left[\frac{L_x}{\lambda}\left(u - m\lambda/p\right)\right] \tag{6.4}$$

where the first terms in parentheses on the right side is the element factor; the second remaining term is the array factor.

For a solid aperture of equivalent aperture length, only the central sinc function would be obtained in a directivity pattern. If a main lobe is examined assuming the element factor there is broad enough to be neglected,

$$L_x\mathrm{sinc}\left(\frac{L_x}{\lambda}u\right) = L_x\mathrm{sinc}\left(\frac{L_x}{\lambda}\sin\theta\right) \tag{6.5}$$

If the same small angle approximation (Fresnel approximation) is made as used for Eq. (5.15), where $\sin\theta \sim \tan\theta$, then

$$L_x\mathrm{sinc}\left(\frac{L_x}{\lambda}\sin\theta\right) = L_x\mathrm{sinc}\left(\frac{L_x x}{\lambda z}\right) \tag{6.6}$$

which is the same expression, Eq. (5.18), used for a solid aperture, where $L_x = np$.

6.3.2 Directivity Simulator

The Directivity Simulator, with its GUI presented in Fig. 6.10, is a graphical implementation of Eq. (6.4). The inputs are p/λ, w/p, number of elements, N, and steering angle, θ. When the array is sampled at the Nyquist spatial frequency rate or better, that is, at $p \leq \lambda/2$, the main beam can be reproduced at the steering angle. For values of p larger than $\lambda/2$, grating lobes may become significant artifacts in the resulting image. The Directivity Simulator in Fig. 6.10 is displaying the case for $p = \lambda/2$, $w/p = 0.8$, $N = 7$, and $\theta = 0$. Grating lobes at the integer values of u are greatly suppressed. This situation is worse for the case shown in Fig. 6.9 which can be reproduced by the simulator for the parameters: $p = \lambda$, $w/p = 0.5$, $N = 7$, and $\theta = 0$.

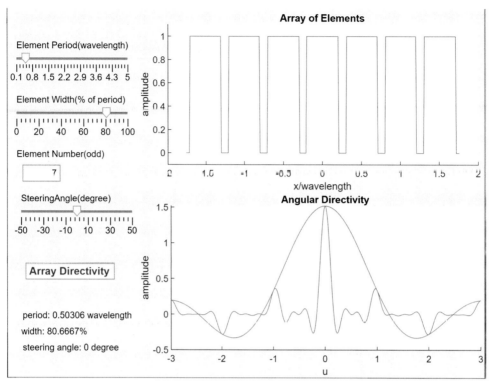

Figure 6.10 Directivity Simulator with input parameters on the left side and output displays on the right.

6.4 Three-dimensional continuous wave array focusing and steering

6.4.1 Three-dimensional beam visualization

With reference to Fig. 6.11, the full 3D beam from a linear array is sketched showing the two different orthogonal focusing mechanisms. Along the x direction, an array of cuboid-shaped elements is electronically focused in the xz imaging plane, or azimuth plane, highlighted in Fig. 5.14. This focusing is accomplished by altering the relative phasing among elements so that the phase front is in the shape of a semicircular electronic lens. Roughly, the beamshape from a well-sampled aperture of length L_x would be approximately like that of a continuous aperture of the same length. If the sampling period exceeds half a wavelength, then grating lobe artifacts may appear. When the beam is steered by adding a linearly increasing phase to each array element, grating lobe effects can be exaggerated.

However, as explained in Section 5.2.4, in the y dimension, each element has an extent on the order of tens of wavelengths and, barring complications from unfavorable vibration modes in the element that are beyond the scope of this course (Szabo, 2014c),

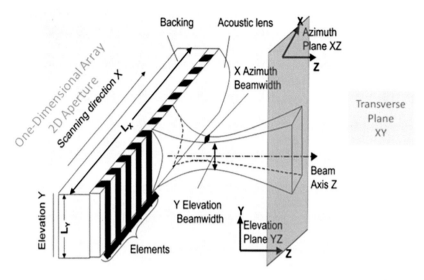

Figure 6.11 Array showing azimuth, elevation, and transverse planes cut planes of a three-dimensional beam (Panda, 1998). *Adapted from Panda, R. K. (1998). Development of novel piezo-electric composites by solid freeform fabrication techniques (Dissertation). New Brunswick, NJ: Rutgers University.*

it radiates in the yz plane as a solid aperture with uniform amplitude. If a physical acoustic lens is placed on the surface of the array, then mechanical focusing will take place at a single location in this elevation plane. If no lens is used, the aperture will radiate a nonfocused beam. These descriptions are simplifications. The real radiated beam is a 3D combination of both focusing mechanisms and therefore is more difficult to visualize. One option for viewing portions of the 3D field is to use a transverse xy plane which can be moved along the z-axis as shown in Fig. 6.11. Another option is to slide the xz plane up and down along the y-axis.

6.4.2 Continuous Wave Array Simulator for 3D beams

The Continuous Wave Array Simulator (CW Array Simulator) has many new features to help explain the nature of a 3D beam compared with the Field Simulator of Chapter 5. As is evident from Fig. 6.12, the input variables have been expanded to cover more than the array variables included in the Directivity Simulator. Another addition is the option to select media other than the lossless case to include the effects of absorption. An alternative output other than pressure and intensity is temperature, a consequence of absorption. Fig. 6.12 shows an intensity plot in the azimuth xz plane for an undersampled array with $p = 2\lambda$, where traces of grating lobes are evident. Finally, because the full 3D beam is calculated, there are new viewing options other than the ones used in the Field Simulator.

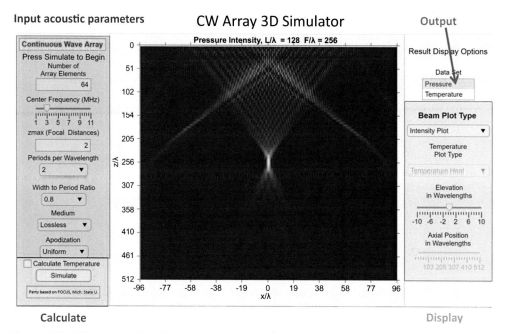

Figure 6.12 CW Array 3D Simulator with inputs, calculations, output, and display.

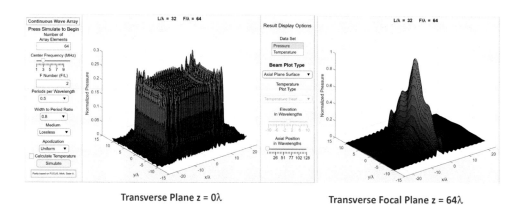

Transverse Plane z = 0λ   Transverse Focal Plane z = 64λ

Figure 6.13 (Left) Acoustic beam from a 32λ × 20λ aperture focused only in azimuth at 64λ in transverse plane at z = 0. (Right) Beam in transverse plane at z = 64λ.

Transverse views defined in Fig. 6.11 reveal the 3D nature of the beam in Fig. 6.13. Here the elevation aperture is 20λ and is not focused and the azimuth aperture is 32λ and the azimuth focus dominates at 64λ.

The computation for this simulator can be thought of as an angular spectrum model, with a sampled aperture, or array. This simulator relies on the (FOCUS, 2023)

software package based on angular spectrum and fast nearfield methods developed by Professor McGough and described in McGough (2004), Chen and McGough (2006), and Zeng and McGough (2008).

6.5 Absorbing media

6.5.1 Array focusing in absorbing media

Even though frequency power law absorption has been covered in Section 3.5, its impact on focusing has not. The major effect of absorption is that there is an amplitude loss with increasing distance from the aperture. Absorption increases with higher frequencies and varies with materials (tissues) (Szabo, 2014d). There may be a minor shift in depth caused by a small change in the phase velocity in different materials. Intensity plots for a beam in a loss-less medium compared to a beam propagating in muscle show differences; in Fig. 6.14, note the relative brightening of prefocal interference maxima and some change in the appearance of the focal region. Each intensity color plot is normalized to the highest value in the field obscuring the fact that the muscle beam is lower overall in amplitude than the lossless beam.

6.5.2 Heating in absorbing media

Heating in tissue and other materials is closely related to material absorption, Szabo (2014e). The steady-state volume rate of heat generation, Q_v, is directly proportional to the acoustic intensity, I, and absorption α at a point,

$$Q_v(x, y, z) = 2\alpha(x, y, z, f_0)I(x, y, z) \tag{6.7}$$

Each material has a thermal conductivity, κ, and its temperature $T(x,y,z,t)$ can be computed from a diffusion equation. Tissues have additional complications: the

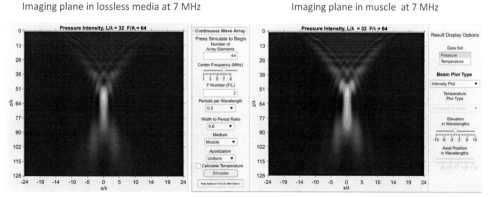

Figure 6.14 CW Array Simulator. (Left) 64 element 3 MHz array radiating into a lossless medium in azimuth plane. (Right) Same array radiating into an absorbing tissue, muscle.

cooling from blood flow which is parameterized by the additional constants: W_b, perfusion rate; C_b, the specific heat of blood; and the temperature of arterial blood, T_a. Overall, the dynamics of heat flow in tissue are described by the steady-state bio-heat transfer equation,

$$K\nabla^2 T - W_b C_b(T - T_a) + Q_v = 0, \tag{6.8}$$

where ∇^2 is the vector Laplacian operator. For materials without perfusion, only the first and last terms of this equation apply. In this case and in tissues with little blood flow, diffusion is the dominant process governing the spatiotemporal evolution of heat distribution. Calculations for temperature elevation in tissues in the CW Array Simulator are carried out by the finite difference approach developed by Zeng and McGough (2005).

6.5.3 Simulation of focusing and heating in absorbing media

The source of heat is the product of local intensity and absorption, as in Eq. (6.7). Because heat diffuses away from the original source of heat, the acoustic beam, the extent of the heated zone is larger than the beam. In general, short pulses at low repetition rates are used in ultrasound imaging, so the time average intensity is much lower than that produced by CW transmission, shown here in Fig. 6.15. On the left is the beam intensity propagating in muscle and on the right, the temperature elevation once equilibrium between heat input and diffusion is established. A nice feature of the simulator setup is that once temperature calculation is selected and temperature is selected for display, different types of 3D display options are available. The temperature surface plot provides quantitative information as illustrated in Fig. 6.16. Also, by selecting pressure, the acoustic beam can be shown for comparison, illustrated by side-by-side displays of the same simulation in Fig. 6.15.

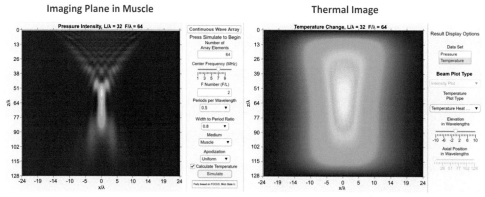

Figure 6.15 (Left) Acoustic beam in azimuth plane from a $32\lambda \times 20\lambda$ aperture focused only in azimuth at 64λ in muscle. (Right) Resulting temperature elevation in muscle.

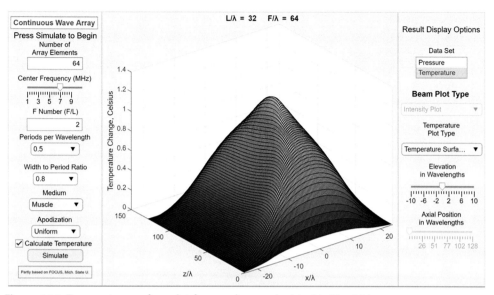

Figure 6.16 Temperature surface plot for muscle case depicted in Fig. 6.14.

6.6 Plane wave compounding

6.6.1 Principles of plane wave compounding

An alternate way of focusing a beam is to stack pairs of steered plane waves, a method called "plane wave compounding" (Montaldo et al., 2009; Szabo, 2014f). Recall that a beam can be steered by applying a linearly increasing phase across array elements, as demonstrated in Fig. 6.3 and depicted by the geometry in Fig. 6.17. Note that the steering is once again affected by the directivity of a single element which has the effect of reducing the amplitude of a steered plane wave with respect to the straight-ahead beam. In principle, only small angles from the vertical are employed so that the sinc-shaped element directivity is not much of a problem.

A straight-ahead plane wave and pairs of steered beams at constant angular increments are summed as depicted in Fig. 6.18 to produce the final beam. Variables are to either change the number of plane waves or change the size of the angular increment. While the advantages of this methodology may not be apparent until Chapter 8, the primary one is an alternative way of focusing. Note that for pulse-echo image formation using scanned beams, a number, N, of pulse-echo lines must be transmitted and received for a total acquisition time of about $N\, 2z/c_0$. Instead, in this relatively new method, only a few plane waves are needed to be transmitted, received, and processed to cover the entire field of view, resulting in about an order of magnitude faster image formation time.

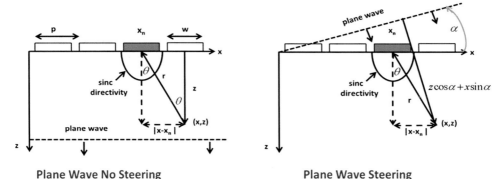

Figure 6.17 Geometry for plane wave steering. (Left) With no steering delays from elements x_n and x_{n+1} to point (x,z) are r/c_0 and z/c_0. (Right) With a steering angle α the delays are $(z \cos \alpha + x \sin \alpha)/c_0$ and r/c_0, where $r = \sqrt{z + (x - x_n)^2}$ (Szabo, 2014f). *From Szabo, T. L. (2014f). Section 10.12.2. In Diagnostic Ultrasound Imaging: Inside Out (2nd ed.). Elsevier.*

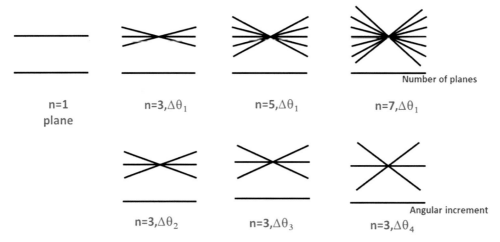

Figure 6.18 Two options for plane wave compounding: (Top) change number of plane waves and keep angular increment the same. (Bottom) Keep the number of plane waves constant and increase the angular increment.

6.6.2 Plane Wave Compounding Simulator

The Plane Wave Compounding Simulator provides ways of experimenting with this approach of beam formation. To facilitate real-time feedback, a computationally efficient continuous wave (single frequency) array algorithm is utilized. So in practice, even though steered plane wave delays are applied to the elements, steered nonfocused diffracting beams are launched instead of ideal plane waves. This kind of complexity adds to the fun of exploration. In addition, the amplitude and phase of each beam are

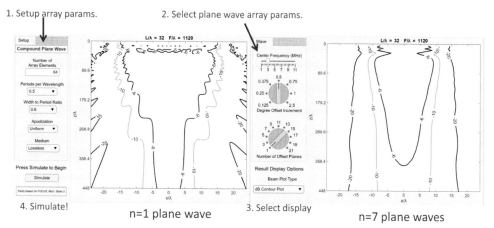

Figure 6.19 Plane Wave Compounding Field Simulator: (Left) ordinary single plane diffracted wave showing a contour display for a 32λ aperture. (Right) Setup steps for simulator and display for seven plane waves. Note a −6 dB contour shows a region of best resolution.

summed over all beams at each field point (x,z). Calculation time depends on the number of plane waves selected.

As can be observed in the Graphical User Interface input panel on the left in Fig. 6.19, the full complement of array variables including apodization, and medium absorption are thrown into the mix. On the right of this simulator are the center frequency, the number of planes, and angular offset, as well as the now familiar display options.

6.7 Lab 6: exploring arrays and continuous wave beams in absorbing media

6.7.1 Exercises with the simulators

This chapter introduces transducer arrays that permit forming beams electronically by applying different signal phases to the individual elements. While most of the chapter describes array beamforming for continuous waves at a single frequency, the first simulator takes a temporal view in which each point source is excited with a short pulse, producing an expanding spherical wavefront much like the rings on the surface of water formed by a tossed pebble. The **Wavefronts Simulator** provides an animated view of the evolution of the wavefronts as they interfere and add together. The first exercise uses a single point source, and the result is very much like watching the rings expand on a pond. When multiple points are used, the student sees that the constructive coalescence of the individual wavefronts occurs very near the array because the source points are spaced only half a wavelength apart. The line distribution of points permits observing the transition from a point source to an extended linear aperture which produces a flat wavefront parallel to the source with spherical behavior at the edges. Steering the beam clearly illustrates the use of signal delays applied to each point source. Finally, observing a focused wavefront propagate shows the initial curvature of the wavefront, needed to focus the energy near the center of curvature. The buildup of wave height can be dramatic as the wavefront approaches the focal region and is an excellent illustration of array gain. Perhaps counterintuitively, the wavefront curvature progressively decreases and becomes flat at the focus. Soon, the waves begin to diverge again, and the focal zone field can be interpreted as a narrow aperture source of its own.

The discussion in Sections 6.2 and 6.3 makes much use of Fourier transforms of well-known functions along with the convolution theorem to explain the role of element directivity and array sampling, leading to the important results plotted in Fig. 6.9 that illustrate array directivity and the presence of grating lobes. The **Directivity Simulator** brings the figure and sampling equations to life by allowing the student to adjust key parameters and observe the results of making practical tradeoffs. For example, element width, pitch, and the number of elements influence the narrowness of the array beam and the strength of sidelobes. The consequences of these choices, while not discussed here, affect the complexity and cost of the imaging system.

The **CW Array 3D Simulator** uses the computational power of the FOCUS package to provide rapid field calculations of 1D arrays with cuboid elements having dimensions in both x and y and operating in continuous wave mode, that is, in steady state at a single frequency, visualized in axial or transverse slices through 3D space. By exploring the field in this way, the student gains further appreciation for the intricacies of the prefocal region, the transition to the focal region, then beyond to the postfocal region. Sidelobes can be observed and quantified, as can grating lobes when the arrays are "undersampled." In addition, the concepts of absorption and heating due to the transfer of energy from the beam to an absorbing medium can be explored quantitatively, illustrating how prefocal sidelobes can gain importance in attenuating media and illustrating the diffusion of heat into a region much larger than the acoustic focus where the acoustic intensity, source of the heating, is highest.

Finally, the idea that a few unfocused beams can be used to provide very fast image frame rates is introduced in Section 6.6 and with the **Plane Wave Compounding Simulator**. Using the simulator, students can see the way that summing plane wave acquisition data for several different steered unfocused beams (linear phase delay transmissions) can effectively produce a focused beam retrospectively. The procedure allows for setting up array parameters, selecting plane wave parameters, and even a tissue medium, then launching the computation; several display methods to visualize the resulting field distribution are available.

6.7.2 Experiments and exercises with the Vantage system

The Vantage control panel shown in Fig. 6.20 now adds several more choices to those already implemented in prior labs: the element separation (array pitch) can now be

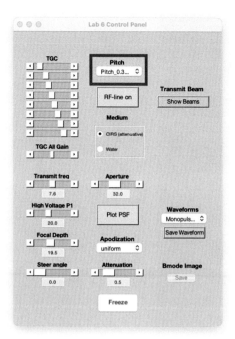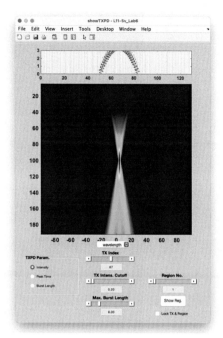

Figure 6.20 (Left) Screenshot of the Vantage GUI control panel for Lab 6. Here, the ability to change the effective array pitch has been added. (Right) Image of the built-in Vantage transmits beam simulation result for a typical focused beam with pitch $p = \lambda$.

modified to observe image artifacts from undersampling the array; the attenuation can be adjusted to compute the simulated field in an absorbing medium and to correct the reconstruction (receive beamforming) to compensate for the loss of signal with depth; the image quality and frame rate of images from plane wave compounding can be compared to those obtained with scanned focused beam acquisition. Delays (phases) applied to the individual elements in the aperture are now displayed along with apodization profiles when plotting the simulated transmit beams so that students can appreciate at a glance how different options affect beam formation.

As presented in Fig. 6.9 and explored with the Directivity Simulator, grating lobes appear when the array is sampled with pitch $p > \lambda/2$. The impact on the transmit (and receive) beams is illustrated in Fig. 6.12 and using the CW Array 3D Simulator, and leads to imaging artifacts in which "ghost" images of targets in the center of the image, produced by the zeroth order beam ($m = 0$), also appear where the first order beam ($m = 1$) is directed. Imaging the FATHOM phantom allows the students to see these artifacts as a function of the array pitch, as illustrated in Fig. 6.21. The yellow dashed lines help identify the ghost images of the point targets from the grating lobes.

The effect of steering the beam can also be observed for the different pitch options and the element directivity plays an important role in the relative strength of

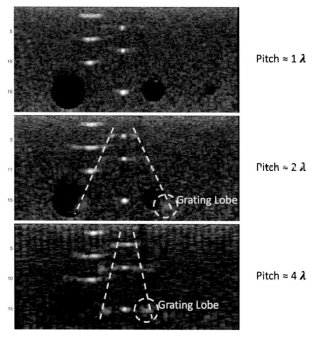

Pitch ≈ 1 λ

Pitch ≈ 2 λ

Grating Lobe

Pitch ≈ 4 λ

Grating Lobe

Figure 6.21 (Top) Image of the FATHOM phantom with no noticeable artifact from grating lobes, given the L11−5 array with a pitch of about 1 wavelength. (Middle) Image of the FATHOM phantom with clearly visible grating lobes, using the L11−5 array with an effective pitch of about two wavelengths between elements. (Bottom) Imaging with a pitch of four wavelengths produces close-in ghost images from the strong grating lobes.

the sidelobes. Array parameters for the L11−5 transducer are provided so that the student can simulate the array used in the lab.

The FATHOM phantom can also be used to estimate the elevation focus of the array because it has a special array of filaments oriented at 45 degree to the usual resolution target filaments. This arrangement permits direct visualization of the elevation width (in the y-dimension) of the imaging region at a number of depths and by looking for the minimum provides a very intuitive way of finding the location of the elevation focus. This measure can be compared to looking for the depth of the bright horizontal band where the backscatter from the speckle is greatest (speckle is introduced in Chapter 8), but this approach is severely complicated by the presence of absorption, as illustrated in Fig. 6.22, right panel.

Pulse lengths and beamplots are measured in several locations using the plotting tool introduced in Chapter 5 to provide quantitative measures of image quality and artifacts to reinforce the observations made using the simulators (e.g., Fig. 6.12) and the Vantage system. In this way, students develop a firm intuitive understanding of the principles underlying arrays and beamforming.

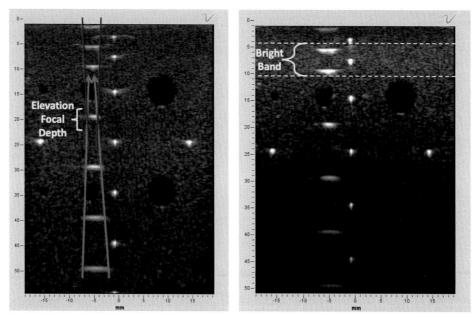

Figure 6.22 (Left) Image of the FATHOM phantom used to visualize the depth of the elevation focus, located where the 45 degrees filaments, located in between the regular filaments (bright points), appear to be shortest: 20 mm is indeed the specified elevation focal depth. (Right) Image of the FATHOM phantom produced using a Plane Wave and with uniform TGC. The bright band indicates a depth zone of increased intensity due to both elevational focusing by the transducer's lens and absorption which biases the estimate of the focal depth significantly!

References

Bracewell, R. N. (2014). *The Fourier Transform and Its Applications* (3rd ed., pp. 82–83). India: McGraw Hill Education.

Chen, D., & McGough, R. J. (2006). A 2D fast near-field method for calculating near-field pressures generated by apodized rectangular pistons. *The Journal of the Acoustical Society of America, 124*, 1526–1537.

FOCUS (Fast Object-Oriented C++ Ultrasound Simulator). (2023). *Features*. Available from https://www.egr.msu.edu/~fultras-web/index.php. Accessed March 13.

McGough, R. J. (2004). Rapid calculations of time-harmonic nearfield pressures produced by rectangular pistons. *The Journal of the Acoustical Society of America, 115*, 1934–1941.

Montaldo, G., Tanter, M., Bercoff, J., Benech, N., & Fink, M. (2009). Coherent plane-wave compounding for very high frame rate ultrasonography and transient elastography. *IEEE Transactions on Ultrasonics, Ferroelectrics, and Frequency Control, 56*, 489–506.

Panda, R. K. (1998). Development of novel piezoelectric composites by solid freeform fabrication techniques (Dissertation). New Brunswick, NJ: Rutgers University.

Szabo, T. L. (2014a). Overview. *Diagnostic Ultrasound Imaging: Inside Out* (2nd ed.). Elsevier.

Szabo, T. L. (2014b). Array beamforming. *Diagnostic Ultrasound Imaging: Inside Out* (2nd ed.). Elsevier.

Szabo, T. L. (2014c). Acoustic wave propagation. *Diagnostic Ultrasound Imaging: Inside Out* (2nd ed.). Elsevier.

Szabo, T. L. (2014d). Attenuation. *Diagnostic Ultrasound Imaging: Inside Out* (2nd ed.). Elsevier.

Szabo, T. L. (2014e). Ultrasound-induced bioeffects. *Diagnostic Ultrasound Imaging: Inside Out.* (2nd ed.). Elsevier.

Szabo, T. L. (2014f). Section 10.12.2. *Diagnostic Ultrasound Imaging: Inside Out*. (2nd ed.). Elsevier.

Zeng, X., & McGough, R. J. (2005). Multiplanar angular spectrum approach for fast simulations of ultrasound therapy arrays. *The Journal of the Acoustical Society of America, 118*, 2967–2977.

Zeng, X., & McGough, R. J. (2008). Evaluation of the angular spectrum approach for simulations of near-field pressures. *The Journal of the Acoustical Society of America, 123*, 68–76.

CHAPTER 7

Pulsed phased array beamforming

7.1 Overview

7.1.1 Pulsed arrays

The same block diagram as that in Fig. 6.1 applies here except that instead of altering the phase to each element, temporal pulses are delayed. In a pulsed array, a selected electronically generated input pulse with a specifically determined delay is sent to each element to form an electronic lens for focusing the transmitted beam. Fig. 7.1 shows the basic principle: a pulse with a delay path length of A_n is selected so that when added to the naturally occurring delay path, B_n, from the element to the focal point, the sum equals the focal path length F (plus a small constant offset which is the same for all paths). Because all the total paths are the same, the pulses add coherently at the focal length F and result in a focal gain equal to N times the individual transmitted identically shaped pulses. An example of a broadband transmitted pulse is shown in Fig. 7.2. As the transmitted pulses race to the focus, the beamwidth decreases and creates a narrower beam in the "focal region," as was illustrated using the Point Source Simulator in Chapter 6. The electronic lens curvature is circular or approximately parabolic in shape. As will be explored later, the shape and length of the pulse at the focal length determines imaging axial resolution as well as the shape of the beam. For a deeper investigation of the topics in this chapter, more information and resources can be found in Szabo (2014a).

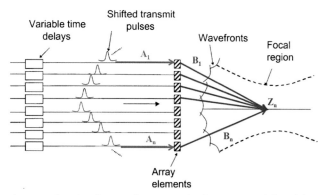

Figure 7.1 Transmit array showing how pulses to each element are delayed by path A_n to add to propagation delay B_n so that pressure pulses are synchronous at the focal length $Z_n = F$ (Panda, 1998). PhD Dissertation, *Development of Novel Piezoelectric Composites by Solid Freeform Fabrication Techniques*. Rutgers University *Adapted from Panda (1998).*

Essentials of Ultrasound Imaging
DOI: https://doi.org/10.1016/B978-0-323-95371-9.00010-2

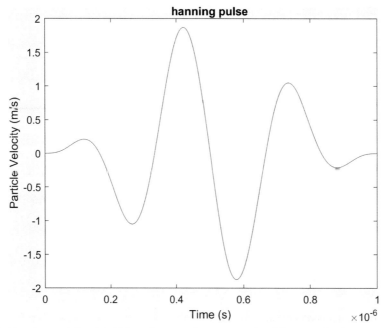

Figure 7.2 Hanning weighted 3 MHz pulse waveform transmitted from each element with appropriate delays.

7.1.2 Wavefront animation simulations

Similar to the earlier thin wavefronts generated by the Point Source Simulator of Section 6.1.2, more complicated wavefronts (ripples), such as the one plotted in Fig. 6.4, are launched from the elements of an array as shown in Fig. 7.3 and 7.4. Here the thickness and shape of the ripple can be changed by differently shaped input waveforms. A time sequence shows the evolution of the wavefronts from a 20-element array in Fig. 7.4.

This propagating ripple has both a lateral extent and an axial extent which can be quantified. Consider the pressure ripple to be described by a function $P(x, t-z/c)$. Have a look at Fig. 7.5. Usually, the envelope of the axial waveform at $x = 0$ is frozen at delay time z/c, and is taken as the response for determining the axial resolution at that position. If the beam is steered (more on this later) then the waveform at the delay distance r/c is taken along the steered direction. The peak amplitude (peak of the pressure envelope) of this waveform becomes the value at one point (here $x = 0$, $t = z/c$) in the lateral beam profile along x at the distance z. To summarize, the pulsed wave one-way (transmit) beamplot is created from the peak pressure amplitudes of individual waveforms each taken at a different position x to collectively trace out a beam profile. For example, in Fig. 7.5, an axial pulse traveling in the z direction,

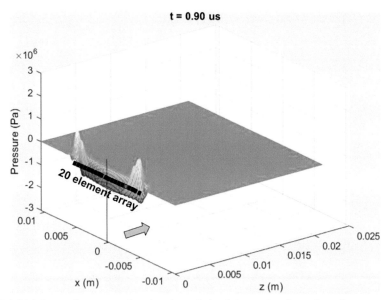

Figure 7.3 Initial frame from an animation from Pulsed Array Simulator for a line array of 20 elements steered at 0 degrees with the input waveform from Fig. 7.2.

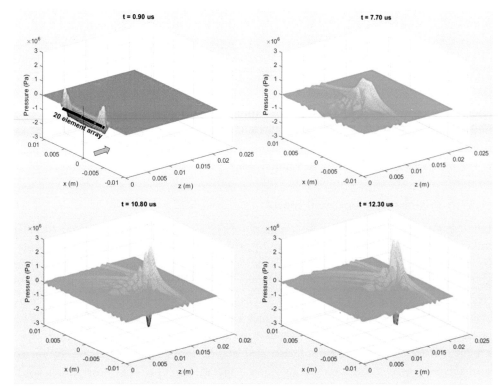

Figure 7.4 Snapshots in time from the Pulsed Array Simulator of a pulsed propagating wavefront focused with F# = 2 steered at 0 degrees for the 20-element linear array in Fig. 7.3.

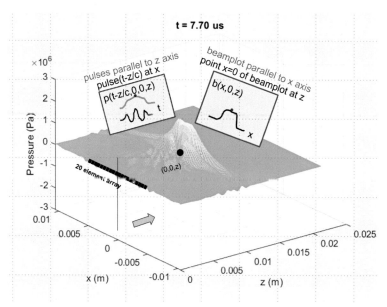

Figure 7.5 A typical transmitted 3 MHz pulse without focusing delay.

frozen at delay $t = z/c$ and associated at a point $(0,0,z)$, is shown as a dot. The peak of the analytic envelope of this waveform becomes the center point value $(b(0,0,z))$ in the beamplot $b(x,0,z)$. The complete beam profile is formed from the neighboring peak amplitudes, all at $t = z/c$, gathered at other values of x. Experimentally, a transmit beamplot is measured in a water tank with a hydrophone which captures each waveform in the ripple at a constant depth z and at each scanned position x. A two-dimensional beamplot in the xz plane requires repeating the x-profile scan over a number of depths. A two-dimensional beamplot in the xy plane is also useful near or in the focal plane. Similarly, a three-dimensional beamplot can be acquired by repeating the measurements over the third dimension.

7.1.3 In this chapter you will learn

Pulsed arrays provide the rapid reconfiguration of scanning beams essential for real time imaging. Pulses add another dimension, that is, axial time, compared to continuous wave arrays. Pulsed waveforms, because of their shape and spectral content, affect not only axial resolution but also beamshape. The electronic beamformer creates the "electronic lens" time delay profile that shapes and focuses the transmit beam. A reciprocal process assembles the pulse-echo signals into a dynamic receive beamformer. The imperfect nature of an aperture in terms of its length and the sampling period of its elements can result in limited resolution and grating lobes. The overall resolution of an imaging system is measured as its "point spread function" in three dimensions.

Absorption in tissues leads to amplitude loss and shape changes in broadband pulsed signals, adding complications that affect achievable resolution. Different types of transducer arrays fall into two major scanning geometries to assemble a 2D image: linear shift and angular rotation.

7.2 How phased arrays form beams

7.2.1 Pulsed array principles

How does a phased array focus? To form an electronically reconfigurable lens for the array to focus in the xz plane as illustrated in Fig. 7.1, a delay for each element of index n is generated for beam formation,

$$\tau_n = \left[r - \sqrt{(x_r - x_n)^2 + z_r^2} \right] / c + t_0, \tag{7.1}$$

where r is the distance from the origin to the focal point, $r = \sqrt{x_r^2 + z_r^2}$, x_n is the distance from the origin to the center of the nth element ($x = np\text{-}p/2$, where p is the array pitch), and t_0 is a constant delay added to avoid negative (physically unrealizable) delays. In other words, the same waveform such as that depicted in Fig. 7.2 is delayed electronically before reaching element n as shown in the overall configuration of Fig. 7.1. As discussed in Section 5.24, the application of a paraxial approximation, that is, the assumption that lateral variations are smaller than the axial distance ($x^2/z \ll 1$), leads to

$$\tau_n \approx (x_n u_s - x_n^2/2z_r)/c + t_0 = \frac{[npu_s - (np)^2/2z_r]}{c} + t_0, \tag{7.2}$$

From this approximate expression, the first term is recognizable as the steering delay with $u_s = \sin\theta_s$ (θ_s is the steering angle) and the second is recognizable as the quadratic phase term needed to cancel the similar term caused by beam diffraction, as shown for a lens in Eq. (5.14). In current practice, the exact Eq. (7.1) is used for arrays rather than its approximation.

The spatial impulse response h of a single element located at position $x_n = np$ is

$$h_n \approx (u, r, t) = a_n h_{0x} w\left(\frac{c}{wu}\right) \prod\left(\frac{t}{wu/c}\right), \tag{7.3}$$

where $u = \sin\theta$, $h_{0x} = \sqrt{c/2\pi r}$ and then the one-way transmit spatial impulse response for an element with focusing can be described as,

$$h_n\left(t - \frac{1}{c}\sqrt{(x - x_n)^2 + z^2} - \tau_n\right) = h_n\left(t - \frac{\sqrt{(x - x_n)^2 + z^2}}{c} - r/c + \frac{\sqrt{(x_r - x_n)^2 + z_r^2}}{c}\right);$$

$$\tag{7.4}$$

at the focus where $x = x_r$, and $z = z_r$,

$$h_n\left(t - \frac{1}{c}\sqrt{(x-x_n)^2 + z^2} - \tau_n\right) = h(t - r/c). \tag{7.5}$$

This is the response h_n of Eq. (7.3) delayed in time. Finally, the overall array response, h_a, is the sum of N elements:

$$h_a(t) = \sum_{-N/2}^{N/2} a_n h_n\left(t - \frac{1}{c}\sqrt{(x-x_n)^2 + z^2} - \tau_n\right), \tag{7.6}$$

which at the focus for $a_n = a$ (unapodized) $h_n = k$ and $r = F$ becomes

$$h_a(t) = Nah(t - F/c), \tag{7.7}$$

with an amplitude gain of N.

Assuming the same shaped waveform, $e(t)$, is injected into each array element, the pressure waveform exiting the element after having been converted to an acoustic pressure pulse by the piezoelectric element at x_n is

$$p_n(x, 0, 0, t) = e(t) * g_T(x_n, t), \tag{7.8}$$

where g_T is the transducer response from the block diagram, Fig. 6.1, and $*$ indicates convolution in time, so that the pressure leaving the transmit beamformer is

$$p_{XB}(x, 0, z, t) = \sum_{-N/2}^{N/2} a_n p_n(x, 0, 0, t) * h_n(x, x_n, 0, z, t, \tau_n). \tag{7.9}$$

Here the beamformer sum is the transmit beamformer represented by block XB (xb in the time domain) in the block diagram. Finally, the pressure at a field point including natural diffraction, h_T, also from the block diagram, and beamformer focusing is

$$p(x, 0, z, t) = p_{XB}(x, 0, z, t) * h_T(x, 0, z, t). \tag{7.10}$$

Another addition from the block diagram is absorption, $a_T(t)$, technically a mirf function from Section 3.5.2 and Eq. (3.24),

$$p(x, 0, z, t) = p_{XB}(x, 0, z, t) * h_T(x, 0, z, t) * a_T(x, 0, z, t, \alpha_0, \gamma). \tag{7.11}$$

This equation embodies the flow sequence of the block diagram, Fig. 6.1 (in time domain form with convolutions instead of multiplications represented in the original frequency domain diagram) from the excitation pulse, $e(t)$, through the transmit beamformer delays $xb(t)$, into pressure pulses transformed by the transducer, $g_T(t)$, naturally diffracted by array as $h_T(t)$, and propagated and absorbed in the selected medium by $a_T(t)$.

7.2.2 Pulsed Array Simulator

The Pulsed Array Simulator offers many more options than the Continuous Wave Simulator because of the added time dimension; in fact, the whole panel is teeming with different input options. Fig. 7.6 shows the sequence of steps for many of the calculations. First, in the excitation signal panel, select a waveform shape (Signal Type), length (Number of Cycles), and frequency (Center Frequency). To simplify the process, the waveform selected as input to the beamformer is really that of a transducer element already excited with an input signal p_n according to Eq. (7.8) (Fig. 7.2 is an example). The array parameters include the pitch spacing in wavelengths, apodization, if any, the width to pitch ratio, w/p, and the number of elements. The $F\#$, steering angle, and the propagation medium are then chosen. Output display parameters are set including the maximum z, the output display parameter, animation, waveform, and its location along the x-axis in wavelengths or the usual options of beamplots or other field displays, and one-way or round trip. Finally, computation is initiated by the "Generate Plot" button.

   This GUI calls on parts of the FOCUS software (2023) to generate different features of the beam. The computations for the Pulsed Array Simulator can be done

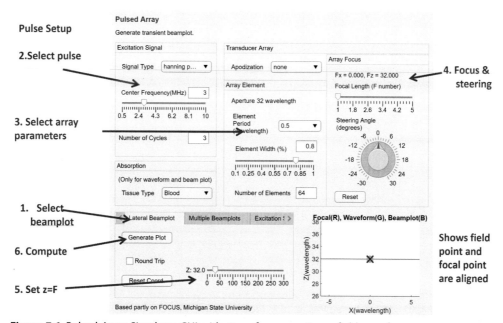

Figure 7.6 Pulsed Array Simulator GUI with steps for generating a field waveform.

rapidly because of the advantages of the FOCUS numerical approach. This method performs transient (time domain) calculations using Time–Space Decomposition and the Fast Nearfield Method (Zhu et al., 2012). More details on how speed and accuracy are achieved can be found in McGough (2004), Kelly and McGough (2006), and Chen and McGough (2006). To speed up computation, the transmit and receive apertures are assumed to be identical. Absorption is added through the methods of Section 3.5. To produce the drive pulse for the array, the equivalent of the pressure in Eq. (7.8) is produced by

$$p_n(r, t) = \rho_0 \partial v / \partial t, \tag{7.12}$$

in which v is the particle velocity normal to the outward face of the array element; one can use either the surface pressure or the normal velocity to describe an acoustic source.

7.3 Effects of pulses and absorption on beams

7.3.1 Array drive pulse effects on beamshape

Altering the drive waveform affects the beam. In general, broadband pulses result in smoother beams with fewer sidelobes. The longer the pulse, the more the beams approach those obtained from continuous wave arrays, as indicated in Fig. 7.7. Here a 64 element array with half-wavelength spacing and an $F\# = 1$ at 32 wavelengths are excited by Hanning-tapered bursts of three and seven cycles. On the left of this

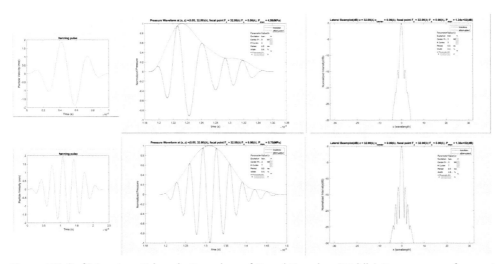

Figure 7.7 (Left) Input particle velocity pulses of 3 and 7 cycles. (Middle) Pressure waveforms at focal point. (Right) Lateral beamplots at the focal distance ($F = 32$ wavelengths) all for a 64-element 3 MHz half-wavelength sampled array.

figure are the particle velocity input waveforms. Pressure pulses launched from the elements are derivatives of these waveforms. In the middle are the pressure signals at the focal point: these are replicas of the initial transmitted pressure waveforms, delayed in time according to Eq. (7.7). From the analytic signal envelope, e the peak values can be found, or $\text{Max}_t\left[\left|e_{\max}(x,t)\right|\right]$ at the selected distance z as described in detail in Section 7.1.2. In this unsteered case, $x = 0$ and $z = F$. Similarly, at another value of x at the same distance z, another peak waveform value can be found. These peak values trace out the lateral beamplot at that distance. For a dB scale, the maximum peak of the beamplot could be used as a reference value. A beamplot in dB (normalized to 0 dB) can be described mathematically for a constant z as

$$I_{\text{dB}}(x) = 20\log_{10}\left(\frac{\text{Max}_t\left[\left|e_{\max}(x,0,z,t)\right|\right]}{\text{Max}_x\left\{\text{Max}_t\left[\left|e_{\max}(x,0,z,t)\right|\right]\right\}}\right). \qquad (7.13)$$

On the right side of Fig. 7.7 are beamplots in dB showing an increase of the number of sidelobes as the pulse is lengthened.

7.3.2 Absorption effects on beamshape

Absorption has an even more profound effect on a pulsed beam than on the beam in the continuous wave case (Szabo, 2014b). Power law absorption increases with both frequency and distance. The mirf which encodes both absorption and dispersion in the time domain (as described in Section 3.5) is convolved in time with the field waveform, Eq. (7.11). If the absorption is weak, as in the case of blood, for a 3 MHz 64 element $F\#$ 1 array as shown in the top row of Fig. 7.8, the effects are slight on both the waveform and beamplot. In a more extreme case for a similar array but operating at 10 MHz into a moderately absorbing medium (e.g., liver), the changes are more severe. The waveform is distorted by both absorption and attendant phase velocity dispersion. The beamplot for liver is lower in amplitude, here normalized to the maximum peak value in a lossless medium. A more global view of these changes can be found in the surface plots for these two tissues found in Fig. 7.9.

7.4 Pulsed grating lobes

7.4.1 Undersampling

Grating lobes from pulsed arrays differ from those from continuous wave arrays in several important ways. With reference to Fig. 7.10 for a CW calculation using the Directivity Simulator, recall that the array factor is an unending sequence of sinc functions (on the assumption of an unapodized array for simplicity, see Fig. 6.6) falling on

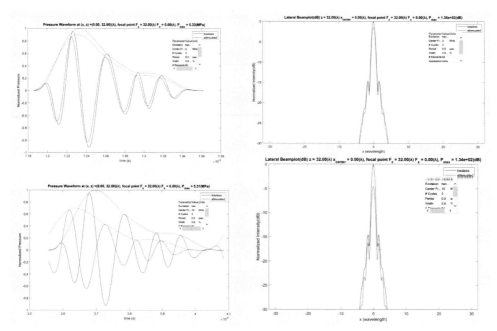

Figure 7.8 Left top: 3 MHz pressure waveform at focal length in blood (red line) compared to the lossless case (blue line). Left bottom: 10 MHz pressure waveform at focal length in liver compared to lossless case. Right top: 3 MHz pressure beamplot at focal length in blood compared to lossless case. Right bottom: 10 MHz pressure beamplot at focal length in liver compared to the lossless case. Note all waveforms and beamplots are normalized to maxima in water.

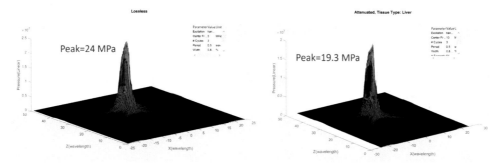

Figure 7.9 Left: Surface plot of pressure field for lossless case for a 64 element 10 MHz *F#* 1 array. Right: same array propagating into absorbing liver. Note the amplitude scales are different.

the grating lobe angles (see Eq. (6.3)), all modulated by an element factor determined by the width of an element (see Eq. (6.4)). For the case shown for a period of two wavelengths, the first grating lobes fall on $\pm 30°$. Because each cycle in a continuous wave is identical, there are many more opportunities for transmissions from different

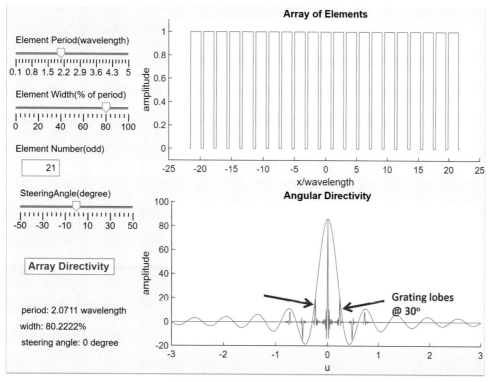

Figure 7.10 Directivity Simulator panel used for a CW twenty-element array with $p = 2\lambda$ and $w/p = 0.8$ and $0°$ steering angle. Relative first grating lobe/peak ratios are ~ 0.2, equivalent to -14 dB.

parts of the array to align and add, thereby creating strong grating lobes. This may not be the case in general, especially for broadband pulses such as those in Fig. 7.8, where few of the cycles are the same. As a result, grating lobes for a broadband array can have the general appearance shown in Fig. 7.11, where they are spread over a larger angular range, are centered at $\pm 22°$, and are lower in amplitude. Specifically, this is a case for the same $p = 2\lambda$ and $N = 20$ elements but with a 3 MHz, three cycle Hanning drive pulse, an $F\# = 1$ focal condition, and the beamplot is viewed in the focal plane.

7.4.2 Steering

As shown in Fig. 7.12, the amplitude of a continuous wave steered beam is limited by the element factor and can have a prominent grating lobe. For the undersampled pulsed case shown previously but this time steered at 20 degrees, the beamplot in Fig. 7.13 is more distributed, has strong grating lobes, and suffers the same overall element factor limitation.

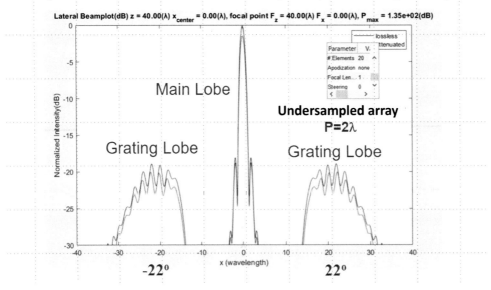

Figure 7.11 Pulsed array simulation of grating lobes for a twenty-element unsteered array with $p = 2\lambda$ and $w/p = 0.8$ and a 3 MHz three cycle Hanning drive pulse. Grating lobe levels are about -20 dB below peak (0.1) and are roughly centered at $\pm 22°$. Blue line is for an unattenuated medium, red line is for blood.

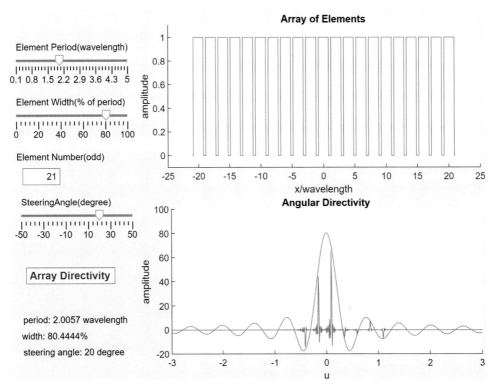

Figure 7.12 Directivity Simulator example: Twenty-element array with $p = 2\lambda$ and $w/p = 0.8$. The continuous wave beam is steered at 20 degrees. The grating lobe is only about 6 dB below the peak of the steered beam.

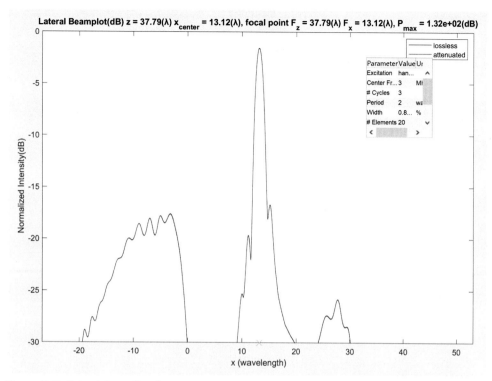

Figure 7.13 Pulsed Array Simulator example showing undersampled case steered at 20 degrees in water. The strongest grating lobe is about 17 dB below the main lobe of the steered beam.

7.5 Combined receive and transmit beamforming

7.5.1 Array receive focusing

Receive beamforming has advantages over transmit beamforming. A conventional transmit beamformer can focus at only one focal point per transmission. A receive beamformer can achieve focusing at all depths, ideally. In order to understand how this is possible, refer to Fig. 7.14. The path of a pulse echo from a point z on axis to element n in an array has a propagation delay τ_{pn}. If a compensating delay τ_{dn} is added to the propagation delay to obtain the same constant overall time for each element channel, then all the pulse-echo signals will be aligned in time and can be added in synchronism. This approach is called a "delay and sum" beamformer. An additional advantage is that the compensating delay profile or electronic lens can adapt for each depth. Depending on how fast the electronic lens can be reconfigured relative to the digitized pulse-echo data streaming into the beamformer determines the number of individual depths at which the refocusing can occur. This process of reconfiguring the

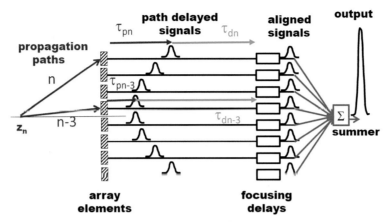

Figure 7.14 Operation of delay and sum beamformer: the path from point z to element n is compensated by an added delay so that all paths have the same delay for coherent summation. Adapted from (Panda, 1998). PhD Dissertation, *Development of Novel Piezoelectric Composites by Solid Freeform Fabrication Techniques*. Rutgers University.

receive focusing with depth is called "dynamic focusing." This focusing is compensating for diffraction which occurs in reverse for reception.

Consider an expanding spherical wave expanding from a point scatter at location $(0,0,z_n)$ as shown on the left side of Fig. 7.14. An element will still have its directivity characteristic, even on reception, so that its directivity values will change depending on the angle between the location of the element relative to the field point location. Similarly, on a larger scale, the directivity of the array itself depends on the placement (and ultimately for other array geometries, on the orientation) of individual elements. Refer once again to Fig. 7.1 for a transmit focusing beamformer. Note that the propagation paths B_n converge on a single point F. Because of the added compensating delay paths A_n, the pulses are synchronous at F. Now replay this configuration in reverse: the scattering point at $(0,0,z)$ replaces $(0,0,F)$; propagation delays to the elements are the same and then compensating delays are added; thus, the end result is a synchronous combination of electrical pulses.

Mathematically, the combination of these focusing and diffraction relationships has a form similar to that for transmit. Consider an incident pressure pulse, $p_{in}(t)$, arriving at a point scatterer located at $(x,0,z)$ and which is scattered or diffracted back to an element n at x_n represented by h_R,

$$p_{RBn}(x - x_n, 0, z, t) = p_{in}(x, 0, z, t) * h_R(x - x_n, 0, z, t). \tag{7.14}$$

The pressure arriving at an element n is converted via the piezoelectric effect, represented by transducer element response, g_R, and delayed appropriately by τ_n

[see Eqs. (7.1) and (7.2)] to make e_n synchronous with the electrical signals from all the other elements,

$$e_n(t) = p_{RBn}(x - x_n, 0, z, t) * g_R(x - x_n, 0, z, t) * \delta(t - \tau_n), \qquad (7.15)$$

and finally, the beamformer sums up the contributions from each element weighted by an apodization coefficient, a_{Rn} for each element (or not when $a_{Rn} = 1$) to obtain the overall receive beamformer response for each depth, $e_{RB}(z,t)$,

$$e_{RB}(z, t) = \sum_{-N/2}^{N/2} a_{Rn} e_n(z, t). \qquad (7.16)$$

7.5.2 Array round trip responses

To complete the picture of the pulse-echo beamforming process, refer to Fig. 7.15. In Fig. 7.15A, a transmit beam propagates as a time-dependent wavefront such as those depicted in Fig. 7.4. An object intercepts this wavefront and scatters it back to the array as in Fig. 7.15B. The wavefront where each element picks up parts of the wavefront as it passes through the array as shown in Fig. 7.15C. Finally, the received element signals are processed by the beamformer, depicted in Fig. 7.15D.

The overall chain of events from transmit to receive, captured in the block diagram of Fig. 6.1, has been described in a slightly rearranged form in the time

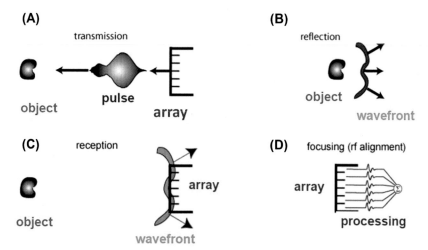

Figure 7.15 The essential sequence of events in pulse-echo processing: (A) transmit pressure pulse released by beamformer, (B) transmitted wavefront is scattered by target object, (C) received wavefront enters array, and (D) received element signals processed by beamformer (Szabo, 2014c). From Szabo, T. L. (2014c). Section 8.4: Role of imaging. In *Diagnostic ultrasound imaging: Inside out* (2nd ed.). Elsevier.

domain. Translated into the frequency domain, without absorption, it can be summarized as

$$E(f)XB(r_n, r, f)G_T(r_n, f)H_T(r_n, r, f)S(r, f)H_R(r_n, r, f)G_R(r_n, f)RB(r_n, r, f), \qquad (7.17)$$

in which the electrical signal, E, is delayed appropriately for focusing by the transmit beamformer, XB (Eq. (7.9)), into the elements where the electrical signals are converted to particle displacements and then pressure waveforms, G_T (Eqs. 7.8 and 7.12). The arrangement of these pressure waves across the aperture causes them to diffract, H_T (Eq. (7.10)) and hit the scattering object with its own directivity characteristic, S. On the return journey to the array, the pressure waves diffract, H_R, and impinge on the elements of the array (Eq. (7.14)), where they are converted to a stream of electrical signals, Eq. (7.15). The receive beamformer then applies the appropriate delays to the returning pulse echoes from each element and assembles them into the resulting beamformed line, Eq. (7.16).

When the scattering object, $S(r, f) = a\delta(r - r_0)$, is a simple point scatterer, and the medium is lossless, a conventional measure of the resolution of an ultrasound imaging system is called the "point spread function." The Point Spread Function (PSF) can be thought of as the imaging version of an impulse response, the spatial pattern for returning echoes from a point target. It can be approximately represented and visualized as an ellipsoid with three axes corresponding to the Cartesian directions as shown in Fig. 7.16. The extent of each axis is determined by the −6 dB amplitude points. Along z, the propagation direction, these points are determined by the half-amplitude width of the excitation waveform envelope, $2a$. Along x, the length is determined by the −6 dB beamwidth in the xz plane (usually the focusing or azimuth plane). Finally, along the y direction, the length of the axes is the elevation beamwidth. The point spread function is not constant across the field of view of the array, and often the array PSF is taken at the depth representing the best overall point spread function, that is,

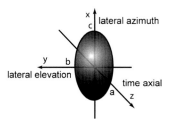

Figure 7.16 The point spread function ellipsoid is constructed from the −6 dB amplitude widths along the coordinate directions by (A) the pulse envelope, (B) the elevation beamwidth, and (C) the azimuth or electronic focusing beamwidth (Szabo, 2014d). From Szabo, T. L. (2014d). Section 7.5.2: Beam-shaping. In *Diagnostic ultrasound imaging: Inside out* (2nd ed.). Elsevier.

the smallest one. This depth is usually where the azimuth or electronic focusing depth is coincident with the elevation focusing depth.

7.5.3 Dynamic receive focusing—sensitivity to sound speed

Beamformers compensate for natural diffraction and focus the beam from the array apertures both on transmit and receive (active aperture widths may be different on transmit and on receive). For pulsed array beamformers, wavefronts of finite temporal duration rippling through a medium are to be visualized rather than a continuous beam extending everywhere. The pulsed array beamformer in the Pulsed Array Simulator has the ability to create a beamplot at a specified distance z from the peak values of individual waveforms as described by Eq. 7.13, a freezing of the wavefront and thus the beamplot in time at z, as described in Section 7.1.2. Therefore the depiction of the overall beam by the Pulsed Array Simulator is a sequence of these frozen beamplots, as in Fig. 7.9 or in any of the available display options such as the stacked beamplots.

For diagnostic ultrasound imaging systems, the round-trip delay to a scattering object at z is taken to be $t = 2z/c$. By accepted convention, c is taken to be the average speed of sound in tissue, $c_0 = 1.54$ mm/μs (Szabo, 2014e). Fortunately, most tissues, because they are mainly water-like in their density, have sound speeds which only vary from the average by a few percent. A worst-case soft tissue is fat which varies from c_0 by 7%. Beamformers are based on the assumption that $c = c_0$ for calculating the delays τ_n in Eq. (7.1). Lab 8 includes observing changes in image quality as the sound speed used for image reconstruction is varied around the true value for the medium. PSFs are notably smaller and brighter when the reconstrution sound speed c matches the true speed c_0 for the phantom; see Fig. 8.33 and Section 8.9.2.

Another common assumption is that the beamformers are operating in a homogeneous medium, one in which the speed of sound is the same everywhere. In reality, the pulses may travel through layers, or other regions in which the sound speed varies, as described in Chapter 2. For example, ultrasound might propagate from a heart wall into blood, or through a layer of abdominal fat and muscle before reaching an organ such as the liver. As explained in Chapter 2, an inclined beam into plane layered media can also undergo a deflection due to refraction. Also, real tissues are often "heterogeneous" which means certain regions may vary spatially (laterally and axially) in their acoustic properties such as sound speed and absorption. Under these circumstances, beamformers function less optimally (Szabo, 2014f). Beamformers working in other applications such as nondestructive evaluation or underwater acoustics where layers having distinctly different speeds of sound can also be compromised in their effectiveness at reconstruction of an image of the medium. In these cases, better results can be obtained by varying the sound speed used in the beamformer to match the average speed of sound in the region of interest. In the cases of delay and sum beamformers which can execute dynamic beamforming by changing the focusing with depth, a further refinement is to change both the focusing and speed of sound with depth to match the imaging configuration. More advanced approaches for dealing with these problems are collectively called "aberration correction" methods, see Szabo (2014f). An implementation of FOCUS including aberration and absorption effects can be found in Szabo et al. (2013).

7.6 Types of arrays

7.6.1 Types of scanning

Conventional approaches to imaging with ultrasound required a means of scanning a beam with adequate resolution across a region of interest. In Section 6.6, one way of achieving scanning employed a series of pairs of tilted "plane waves", and is called plane wave compounding. The most common types of scanning for delay and sum beamformers are depicted in Fig. 7.17. Several elements of a linear array along the x axis, here centered at $x_c = (x_n + x_{n+1})/2$ and highlighted in red in the top left of Fig. 7.17, form an active aperture N_A elements wide that sends a focused beam down centered at this position along the z-axis. Once time is allowed for this beam to reach the selected "scan

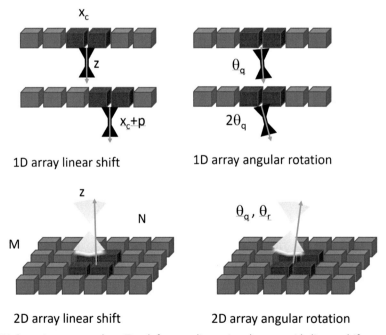

Figure 7.17 Scanning approaches: Top left: one dimensional array with linear shift scanning, Top right: one dimensional array with angular rotation scanning, and Bottom: two dimensional matrix array with linear shift and angular rotation scanning.

depth," the pulse-echo time required to reach a target at the scan depth and return, $t = 2z_{SD}/c_0$, the active aperture is shifted by one element, effectively moving the center of the beam by that amount to center position $x_c = (x_{n+1} + x_{n+2})/2$ as shown in bottom of the top left section of Fig. 7.17. Beam degradation occurs at the ends of the array as fewer and fewer elements are used in the active aperture, as is observed in Lab 7 (Fig. 7.22).

In general, an even number of individually addressable elements are used; the beam is centered on this active subarray. The shift shown here for an array with pitch of $p = \lambda$ is equivalent to one period; however, several elements may be tied together with a single electrical connection forming a group. For example, if the individual elements are spaced at $\lambda/2$, and four elements are combined in a group, the pitch $p = 2\lambda$; then the array is undersampled and the element factor directivity is narrower than in the well sampled case. In this grouping the same delay is given to all the elements in a group and the overall width for the group is the sum of the individual widths. An advantage of this method is that fewer electrical connections or delay channels are required. The beam is directed straight ahead along the z-axis and the grating lobes are off to the side and are weakened by the element factor as shown in Fig. 7.10 and Fig. 7.11. The deteriorated beamplot may also have higher near-in sidelobes due to "quantization" effects; refer to Szabo (2014g) for more information.

Alternatively, scanning can be accomplished by beam steering, employing angular rotation, as shown in the upper right of Fig. 7.17. The great advantage of this approach is only a small active acoustic aperture is required compared to a linear array, and consequently, a reduced acoustic window contact area still provides access to a large internal imaging volume as will be explained in Chapter 8.

In this approach, the beam is scanned through small angular increments, $\Delta\theta$, while the active aperture is held fixed at the middle of the array. Of course, the aperture can be made larger or smaller by changing the number of elements. Once again, scan lines are sequenced at intervals of round trip scan depth times. To distinguish the two arrays which look physically similar in Fig. 7.17, the one on the left is called a "linear array" while the one on the right is known as a "phased array" (Szabo & Lewin, 2013). Furthermore, the linear array has a scan geometry in which the field of view is a rectangular region directly below the array. The phased array has a sector-shaped field of view which extends beyond the edges of the array.

It is important to note that the steered beam is very similar in characteristics to one radiated straight ahead (zero degrees) as shown for examples in Fig. 7.18, focus at 40λ, $N = 20$, $L = 40\lambda$, $F\#1$, $p = 2.0\lambda$, for the pressure waveform at the center of a non-steered beam compared to one at the center of a beam steered at an angle of 20 degrees are very similar in shape; however, the amplitude of the steered beam has dropped significantly by about -2.8 dB and the beamplot is slightly lopsided because of the element factor and has a lurking grating lobe as demonstrated by Fig. 7.13.

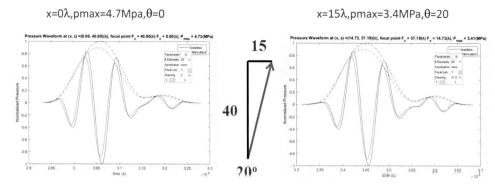

Figure 7.18 Pressure waveforms in water at the center of two beams, one nonsteered on the left and another beam steered at 20 degrees on the right for an undersampled array, $p = 2\lambda$ with $N = 20$. Blue line is for an unattenuated medium, red line is for blood. Note that p_{max} is greater in the unsteered case.

7.6.2 Two-dimensional arrays

At the bottom of Fig. 7.17 are two-dimensional arrays which have $M \times N$ elements. While a one-dimensional array may typically have 128 or 256 elements, such a number would require 16,384 or 65,536 elements to maintain similar size acoustic apertures in two dimensions (Szabo, 2014h). Supplying enough electronic channels for two-dimensional arrays is a daunting task and has been discussed in much more detail in Szabo (2014i). As a result, present practical conventional two-dimensional arrays are $32 \times 32 = 1024$ elements. Approaches to drive two-dimensional matrix arrays with fewer channels than the total number of elements include sparsely connected arrays (more noise, less focal gain), undersampling, and clever beamformers. Unfortunately, undersampling in two dimensions has more severe consequences compared to one-dimensional arrays by creating multiple grating lobes as illustrated by Fig. 7.19, and Turnbull (1991) and Turnbull and Foster (1991).

The main advantage of a two-dimensional array is that it can be electronically focused at a single focal length or point in three-dimensional space, therefore producing the smallest diffraction limited point spread function given large apertures in both x and y dimensions. As shown in the bottom of Fig. 7.17, once a two-dimensional active aperture is selected, either a linear shifting or angular rotation scanning method can be used.

Another approach to reducing the number of electrical connections is called the "row–column method" (Morton & Lockwood, 2003). As shown in Fig. 7.20, a two-dimensional array is constructed from two overlapping elongated orthogonal arrays. Transmission is either on rows or columns and reception is on the orthogonal array. Since there are M rows oriented along the x-axis and N rows along y, each with an electrical connection, there are a total of only $M + N$ electrical connections. Beam coverage is restricted to the two-dimensional area of array and its projection into the medium and it is really the overlapping of the transmitted beam and receive beam that

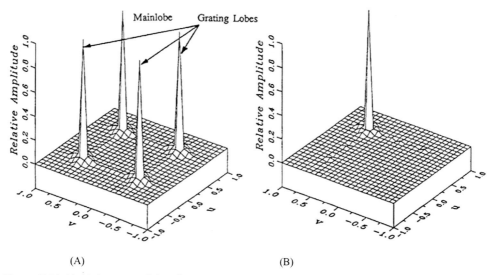

Figure 7.19 Multiple grating lobes for a two-dimensional undersampled array with (A) $p = 1.0\lambda$ and (B) $p = 0.5\lambda$. *From Turnbull, D. H. (1991). Two-dimensional transducer arrays for medical ultrasound imaging (PhD thesis). Toronto, Canada: Department of Medical Biophysics, University of Toronto.*

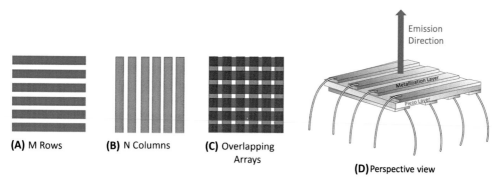

Figure 7.20 Row column configuration: (A) *M* rows of elongated elements. (*B*) *N* columns of elongated elements. (C) Overlap of arrays. (D) Perspective view of the principle. Matching and backing layers not shown.

form the image rather than the geometric overlapping of the apertures. Several options for configuring the arrays including plane and plane wave compounding and others have been devised for improving image quality (Hansen–Shearer et al., 2021).

7.7 Lab 7: Pulsed array investigations

With the introduction to pulsed array beamforming concluded, the students now have a very good appreciation for the process of ultrasound data acquisition, and in

Chapter 8 the complete view of image formation will be added. The following laboratory exercises further reinforce many of the concepts through the use of the Pulsed Array Simulator and its many options, and through the imaging of the tissue-mimicking phantom with the Vantage™ Research Ultrasound System, as described below.

7.7.1 Experiments and exercises with the simulator

The **Pulsed Array Simulator**, with front panel displayed in Fig. 7.6, has many options for configuring the transmission of a short ripple, or wave packet, from an array. The simulator was used to produce many of the figures in this chapter, such as those in Figs. 7.7, 7.8 and 7.9, and the students will be guided to generate similar images through the study of interesting cases. The general approach they will take is to vary the input parameters and use beamplots to inspect the resulting field properties using basic metrics such as resolution and sidelobe strength, and to compare transmit configurations with one another. Exploration of the role of aperture and apodization on field metrics will be familiar from analysis of CW simulations, but deeply understanding the concept of a beamplot for a fleeting wave packet takes some practice. Similarly, developing an intuition for selecting drive waveforms, focusing, apodizing, and steering of the beam, and the effects of grating lobes are important for study.

7.7.2 Experiments and exercises with the Vantage system

The Vantage interface has once again been augmented for this lab, with the addition of a spectral ratio measurement tool, and imaging with a single plane wave instead of a scanned focused beam. Fig. 7.21 presents the Lab 7 front panel.

The Vantage transmit beam plotting tool has already been introduced in Fig 6.20, and to make room for other capabilities, the pitch control has been removed from the Lab 7 interface. The Lab 6 panel can be used to study the effects of altering array periodicity on grating lobes and steering with different pulse shapes, and either program can be used to examine the frequency dependence of grating lobes for the array, since the pitch is approximately one wavelength at the center frequency of 7.6 MHz. The panel for Lab 7 can be applied to examine the roles of apodization, steering, focal delays (focal depth or *F#* choices), and finally, to emphasize the edge regions at the ends of the array. As the active aperture shifts toward the ends of the array, fewer elements are available and the image degrades. The consequences of poor beam shape on image quality depends on the beam itself, and also on the receive beam and where the image reconstruction will be computed. Though this topic will be covered extensively in Chapter 8, the students will be alerted to the "whole system" view and the challenges of imaging at the edges in this lab. They will observe image quality evaluated

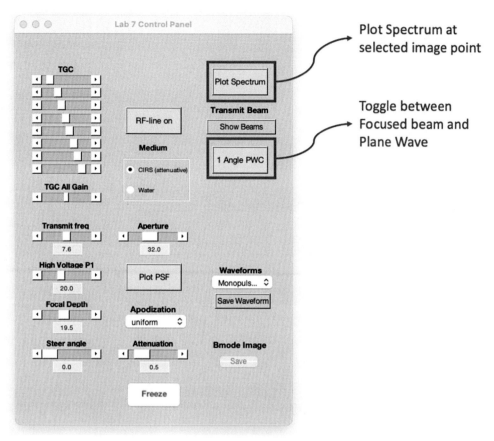

Figure 7.21 Front panel for the Lab 7 GUI, identifying the new controls: Spectral plotting using the RF line through a user-selected point, and an option to toggle between focused beam and plane wave sonication.

by moving wire targets from the middle of the field of view to the edge, as illustrated in Fig. 7.22.

In Fig. 7.22, the images of the phantom's wire targets (appearing as bright points) are presented along with the transmit beams that are used to insonify the region. On the left panel, one can see how the truncated aperture results in a beam that looks like it was steered and somewhat defocused. It is no longer aligned with the location of the beamformed line, visible as a hashed overlay. The receive beamforming still produces a target response, but the PSF is broader and dimmer than the corresponding result away from the edge, as shown on the right. The magnified views (in blue and orange boxes) better illustrate the degradation of the image at the edges. A system designer can compensate for this effect to some extent, but this example illustrates the need to examine such sources of distortion in detail to mitigate the artifact.

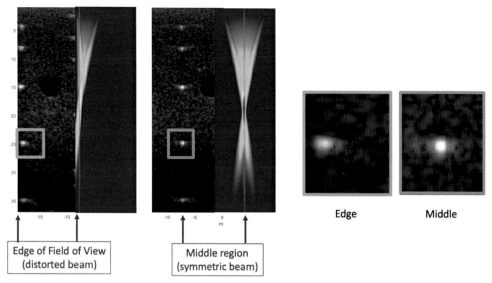

Figure 7.22 (Left) Wire target image on the left edge of the field of view, and a (offset) beamplot of the transmit beam. (Right) Wire target image in the middle region of the field of view, with associated beamplot. (Far Right) Magnified view of the wire targets at the edge and in the middle.

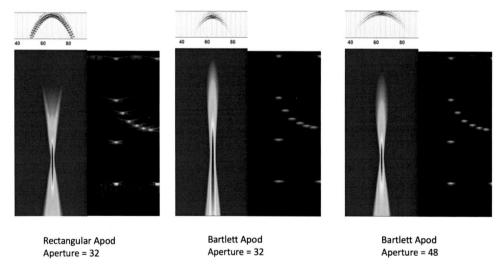

Figure 7.23 Transmit beams and images with rectangular apodization and with Bartlett apodization. Apodized beams can use more elements (right) to approach the beam pattern achieved using fewer elements and a rectangular aperture (left).

Basic features of pulse-echo imaging are worth exploring since the Vantage research system provides easy control over many of the properties explored with the Pulsed Array Simulator. Beginning with apodization and the impact on beam shapes and the PSF, the students can explore conditions such as those illustrated in Fig. 7.23.

Students will easily observe the differences and correlate them with the shading imposed by the apodization function as illustrated by the plots above the beamplot images. Those plots depict both focusing time delays and the amplitude variation of the individual excitation waveforms. The PSF plotting tool gives them a quantitative method to use in characterizing imaging performance. And in response to a wider (lateral) PSF, students can experiment with increasing the aperture size, or changing the transmit frequency.

One can make several interesting observations by changing the frequency and bandwidth of the drive signal. First, lateral resolution is essentially proportional to the frequency, while axial resolution is inversely proportional to signal length, and this can be verified very easily using the imaging phantom, the signal choices available, and the PSF tool. Additionally, the transition between pulsed array behavior and continuous signals can be observed by comparing imaging results using very short pulses (~ 1 cycle) and pulses that are longer than one would normally use for imaging (~ 6 cycles).

But, as was mentioned earlier, one can also examine grating lobes by holding the geometry of the array constant while changing the signal center frequency and examining the location and strength of grating lobes. Predictions using the Pulsed Array Simulator can be directly compared to imaging results. Because grating lobes are often rather weak, they may get lost in a speckle field, but the small custom acoustic phantom (SCAP) refractor insert provides two small pins of submillimeter diameter to image in a free field. Sidelobes and grating lobes are much easier to see in a single scatterer environment. The idea is to observe the grating lobes as the wavelength varies with respect to a constant pitch. See Fig. 7.24 for an example, showing the location and strength of grating lobes at low, mid, and high frequencies within the transducer

Grating Lobes for the L11-5v Array

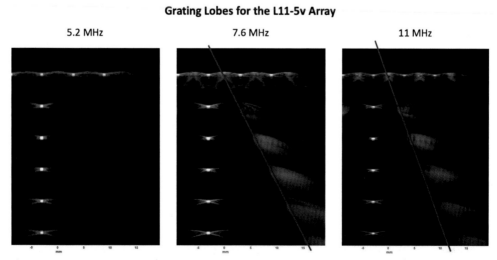

Figure 7.24 Grating lobes observed by varying the frequency for the L11—5v array, when scattering from well-spaced pin targets in water (simulated). Left to right, the effective pitch is 1.0 λ, 1.5 λ and 2.2 λ.

passband. The red lines are a visual guide to the boundaries of the ghost images of the wire scatterers. Near 5 MHz, these grating lobe artifacts are indistinct, and a guide line was too difficult to place. The steeper the guide line, the closer and more prominent are the grating lobes.

Two properties of tissues that interfere with imaging and cannot be avoided are absorption and scattering. While scattering redistributes and redirects ultrasound, absorption converts acoustic energy to heat. Both mechanisms contribute to attenuation of coherent energy in the transmit beam as it propagates to the region of interest, and in the backscattered echo as it propagates back to the array. Measurement of attenuation is straightforward if one has a sample of the acoustic medium and can make a transmission measurement through the sample and compare it with the same result through a reference medium such as water. The problem is more difficult when done in situ, that is, using a backscattering approach; the general idea is that a calibration experiment with a reference medium allows one to compensate for beam diffraction effects (Bigelow et al., 2005). Here, we will use a backscatter reflection measurement to obtain both the sound speed and the absorption of a fluid using the Vantage and the L11−5v transducer. A small tank with a flat bottom is filled with water and imaged with the transducer; the RF data from the two-way path is compared with the same experiment performed through an attenuating fluid with "unknown" properties. The students can obtain the sound speed and the frequency-dependent absorption of the unknown liquid by comparing signals and their spectra obtained from the RF data collected from the reflector at the bottom of the small container. The sound speed is obtained from the relative travel times, given that the speed of sound in water is well known. In general, the absorption is not proportional to frequency but can usually be represented by a power law dependence, as explained in 3.5.1, and the frequency exponent is a characteristic of the medium. Example results, in which the exponent $m = 1.8$, are presented in Fig. 7.25.

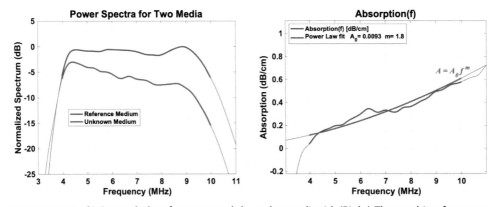

Figure 7.25 (Left) Spectral plots for water and the unknown liquid. (Right) The resulting frequency-dependent absorption.

References

Bigelow, T. A., Oelze, M. L., & O'Brien, W. D. (2005). Estimation of total attenuation and scatterer size from backscattered ultrasound waveforms. *The Journal of the Acoustical Society of America, 117,* 1431–1439.

Chen, D., & McGough, R. J. (2006). A 2D fast near-field method for calculating near-field pressures generated by apodized rectangular pistons. *Journal of the Acoustical Society of America, 124,* 1526–1537.

FOCUS (Fast Object-Oriented C++ Ultrasound Simulator). (2023). https://www.egr.msu.edu/~fultras-web/index.php. Accessed March 13, 2023.

Hansen-Shearer, J., Lerendegui, M., Toulemonde, M., Tang, M., et al. (2021). Ultrafast 3D ultrasound imaging using row-column array specific frame-multiply-and-sum beamforming. *IEEE Transactions on Ultrasonics, Ferroelectrics, and Frequency Control, 69,* 480–488.

Kelly, J. F., & McGough, R. J. (2006). A time-space decomposition method for calculating the nearfield pressure generated by a pulsed circular piston. *IEEE Transactions on Ultrasonics, Ferroelectrics, and Frequency Control, l53,* 1150–1159.

McGough, R. J. (2004). Rapid calculations of time-harmonic nearfield pressures produced by rectangular pistons. *The Journal of the Acoustical Society of America, 115,* 1934–1941.

Morton, C. E., & Lockwood, G. R. (2003). Theoretical assessment of a crossed electrode 2-D array for 3-D imaging. In *IEEE ultrasonics symposium* (Vol. 1, pp. 968–971).

Panda, R. K. (1998). *PhD Dissertation. Development of Novel Piezoelectric Composites by Solid Freeform Fabrication Techniques.* Rutgers University.

Szabo, T. L. (2014a). *Array beamforming, Chapter 7. Diagnostic ultrasound imaging: Inside out.* (2nd ed.). Oxford, UK: Elsevier.

Szabo, T. L. (2014b). *Attenuation, Chapter 4. Diagnostic ultrasound imaging: Inside out.* (2nd ed.). Oxford, UK: Elsevier.

Szabo, T. L. (2014c). *Section 8.4: Role of imaging, Chapter 8. Diagnostic ultrasound imaging: Inside out.* (2nd ed.). Oxford, UK: Elsevier.

Szabo, T. L. (2014d). *Section 7.5.2: Beam-shaping, Chapter 7. Diagnostic ultrasound imaging: Inside out.* (2nd ed.). Oxford, UK: Elsevier.

Szabo, T. L. (2014e). *Introduction, Chapter 1. Diagnostic ultrasound imaging: Inside out.* (2nd ed.). Oxford, UK: Elsevier.

Szabo, T. L. (2014f). *Section 9.7: Aberration correction, Chapter 9. Diagnostic ultrasound imaging: Inside out.* (2nd ed.). Oxford, UK: Elsevier.

Szabo, T. L. (2014g). *Section 7.9: Nonideal array performance, Chapter 7. Diagnostic ultrasound imaging: Inside out.* (2nd ed.). Oxford, UK: Elsevier.

Szabo, T. L. (2014h). *Section 7.6: Two-dimensional arrays, Chapter 7. Diagnostic ultrasound imaging: Inside out.* (2nd ed.). Oxford, UK: Elsevier.

Szabo, T. L. (2014i). *Section 10.12: Alternate imaging system architectures, Chapter 10. Diagnostic ultrasound imaging: Inside out.* (2nd ed.). Oxford, UK: Elsevier.

Szabo, T. L., & Lewin, P. A. (2013). Ultrasound transducer selection in clinical imaging practice. *Journal of Ultrasound in Medicine, 32,* 573–582.

Szabo, T., Nariyoshi, P., & McGough, R. (2013). Acoustic beam simulator with aberration, power law absorption, and reverberation effects. In *2013 IEEE ultrasonics symposium proceedings* (pp. 374–377). Prague, Czech Republic.

Turnbull, D. H. (1991). Two-dimensional transducer arrays for medical ultrasound imaging (PhD thesis). Toronto, Canada: Department of Medical Biophysics, University of Toronto.

Turnbull, D. H., & Foster, S. F. (1991). Beam steering with pulsed two-dimensional transducer arrays. *IEEE Transactions on Ultrasonics, Ferroelectrics, and Frequency Control, 38,* 320–333.

Zhu, Y., Szabo, T. L., & McGough, R. J. (2012). A comparison of ultrasound image simulations with FOCUS and Field II. *2012 IEEE Ultrasonics Symposium Proceedings,* 1694–1697.

CHAPTER 8

Ultrasound imaging systems and display

8.1 Overview

8.1.1 Block diagram

This chapter concentrates on what leads up to forming images and video sequences and the information hidden within them. The back end, highlighted as the last step in the overall signal chain of the central block diagram in Fig. 8.1, includes both signal and image processing functions.

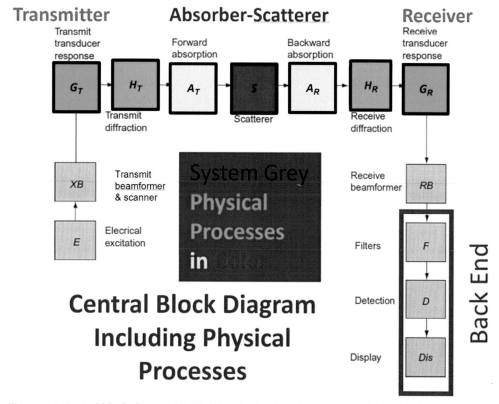

Figure 8.1 Central block diagram highlighting the back-end processing which includes various processing steps and image formation. *Adapted from Szabo, T. L. (2014a). Diagnostic ultrasound imaging: Inside out (2nd ed.). Elsevier.*

Essentials of Ultrasound Imaging
DOI: https://doi.org/10.1016/B978-0-323-95371-9.00009-6

8.1.2 Back end processing

Even though the conventional processing for image formation was well known, a considerable amount of additional processing was also carried out in a typical ultrasound backend (Szabo, 2014b). An on-board computer resident in the imaging system played a dominant role in signal processing functions and for the replacement of special processing hardware.

In the top of Fig. 8.2 are the conventional steps needed for image construction. Radio frequency (RF) pulse echo data streams in and is converted to envelope data for each beamformed line and passes through logarithmic amplifiers to preserve the wide dynamic range of the echo levels and finally, it is preprocessed and converted to digital form. The acoustic lines of data are rearranged in their original geometric layout and are interpolated by a process called "scan conversion" which converts the envelope data to an $m \times n$ matrix of pixels (Szabo, 2014b). An additional step called postprocessing alters the relationship between the image data and their presentation in terms of gray scale available in the monitor. In earlier days, the display monitors were analog so an additional digital to analog conversion step was needed.

In modern systems, these steps are hidden and combined with other information and other options for image presentation, so in the lower part of Fig. 8.2, more generic signal and image processing blocks represent the overall processing. A good example of this latter representation is color Doppler imaging. Here the regular B-mode gray scale image, processed in the traditional way, is overlaid with a visualization of blood flow presented in vivid colors representing both blood velocity and its direction (more detailed explanations follow in Chapter 9). To make this type of image, a considerable amount of signal processing is required to extract Doppler flow information from specialized multiple RF data lines as well as image processing to encode and

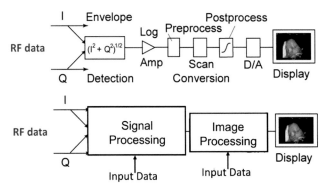

Figure 8.2 Top row: Conventional steps in image formation. Bottom row: A higher level breakout of signal and image processing functions in the back end. *Adapted from Szabo, T. L. (2014a). Diagnostic ultrasound imaging: Inside out (2nd ed.). Elsevier.*

display the Doppler information. Alternative image processing is explained in Section 8.8 and different imaging system architectures and processing are discussed in Chapter 10.

8.1.3 In this chapter you will learn

RF beamformed acoustic lines are transformed into images having different formats. How these images are formed and processed is explained. Hidden inside images are characteristics of the acoustic beams and a measure of image resolution called the "point spread function" (PSF). Phantoms with wire "point" targets provide signatures of beam characteristics as indicated by the Scatter Image Simulator. Ways of extending the lateral resolution throughout the image can be accomplished by multiple focus zones as demonstrated by the MultiFocus Simulator. Because absorption diminishes pulse echo returns with depth, time gain compensation (TGC) is employed. Advantages and shortcomings of TGC are observable through the TGC Simulator. Different types of scatterers visible in images of tissues and organs can be classified by their size relative to a wavelength. Subresolution size scattering appears in the form of speckle which is dependent on the frequency of the ultrasound and beam formation even though it is often mistaken for tissue structure. Examples of speckle are given by the Speckle Simulator. Image quality is measured both by resolution and image contrast. Ways to improve and measure contrast are explained. Finally, several ways of creating ultrasound videos are compared.

8.2 Image formation
8.2.1 Scanning and image formats

The three transducer arrays in most common use today are the linear array, the phased array, and the curved linear or convex array as illustrated in Fig. 8.3 (Szabo, 2014b; Szabo & Lewin, 2013). Each type of array utilizes a different kind of scanning method to cover the image region with lines, two of which were introduced in Section 7.6.1. With reference to the top of Fig. 8.3, an active aperture of n elements is formed to create a beam directed along z. This aperture is shifted by linear translation along x of one period, $\Delta x = p$, at a time, from point a to b in Fig. 8.3. A linear array typically used for imaging vascular and superficial structures is shown at the top right and it produces a rectangular image format (scan shape) shown at top left. In the middle of Fig. 8.3 is a phased array which scans by angular rotation or steering by an active aperture centered on the physical array and increments scanlines by $\Delta\theta$. Typically this scan results in a pie-slice sector shape shown at the middle left, usually to ± 45 degrees. An advantage of this kind of scan and transducer is a small footprint (i.e., acoustic contact window) ideal for cardiac imaging which often requires scanning between the

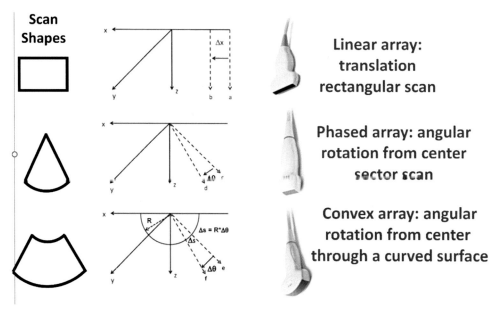

Figure 8.3 Three common methods of scanning. Top: Linear array linear translation scanning for a rectangular format. Middle: Phased array angular rotation scanning for a sector format. Bottom: Convex array for angular rotation scanning for a clipped sector format.

ribs. At the bottom of Fig. 8.3 is a convex or curved linear array in which the elements are on a curved surface with radius R. Scanlines sent perpendicular to this surface are sent out at an angle based on their curved geometry. The increment between scanlines is in terms of an arc length or angle, $\Delta s = R\Delta\theta$. The active aperture is moved along like that of a linear array except on a curved surface. The angular spread of the acoustic lines is caused by the geometric arc. The resulting format is like a clipped pie sector format shown at the botom left of Fig. 8.3.

8.2.2 Image frame construction

In order to understand the composition of a typical image frame by the delay and sum beamforming method (Szabo, 2014b), begin with an individual scan line, $T_s = 2s_d/c_0$ where s_d is the scan depth to the bottom of the image. To fill out a frame, N image lines are shot sequentially for a total time of a frame:

$$T_F = 2Ns_d/c_0. \tag{8.1}$$

From Fig. 8.4, it is evident that certain electrical signals mark the beginning of specific events. The start-of-frame pulse initiates the start of frames ideally at intervals of T_F, though some extra time is sometimes added to make frame rates occur in nice integer numbers per second. The start-of-transmit signal initiates each set of delayed

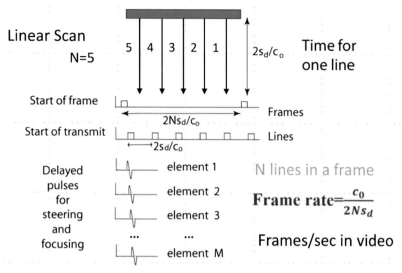

Figure 8.4 Deconstruction of a ultrasound image frame. Beams are launched at intervals along the array. When *N* lines are completed, a frame is filled. The total round trip transit times to the bottom of the frame determine the time a full frame takes. Frame rate is the reciprocal of this time. Shown are the pulses initiating a frame and those starting a line as well as a set of pulses going to each element per line Szabo (2014a). *Adapted from Szabo, T. L. (2014a). Diagnostic ultrasound imaging: Inside out (2nd ed.). Elsevier.*

pulses sent to every element in an active aperture for focusing or steering, etc., and they occur at intervals of T_s. Shown below in this figure are the delayed pulses going to the individual elements that are turned on for an active aperture. The frame rate is simply $1/T_F$. Other frame construction and combined beamforming methods will be covered in Section 8.8.

8.2.3 A matrix of pixels

The lines of Fig. 8.4 are assembled into a picture by a process of scan conversion (Szabo, 2014b). The acoustic lines are turned into amplitude versus time lines (A-lines!) or image lines in their proper geometric arrangement (parallel lines in this case) and then interpolated into a matrix of *m* by *n* pixels (Fig. 8.5). A large dynamic range of data amplitudes are logarithmically scaled (called "compression") and then converted into gray levels (here, 256 values) through a nonlinear postprocessing curve which is selected to emphasize important features or regions of the image (such as enhanced or decreased contrast). The gray scale conversion curve is often shown as a color bar in the image as in Fig. 8.6. The amplitude to greyscale levels conversion curve is usually not displayed, but is shown for the colorbar of this image at the right of Fig. 8.6.

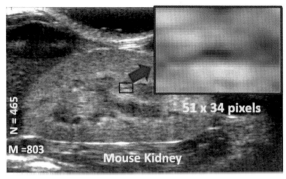

Figure 8.5 Linear array image of a mouse kidney of 803 by 465 pixels. A magnified view of 51 by 34 pixels to show individual pixels.

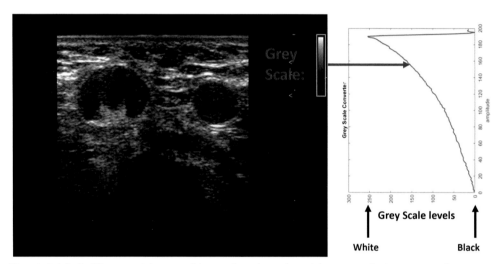

Figure 8.6 Left: Ultrasound image of breast lymph nodes with gray scale in upper right corner. Right: Gray scale conversion map for image: amplitude values on bottom horizontal axis are mapped into unique gray scale values on vertical axis in a nonlinear way.

8.3 Acoustic line adventures

8.3.1 Point spread function ellipsoid revisited

As encountered in Section 7.5.2, the PSF ellipsoid is the -6 dB resolution shape depicted in Fig. 8.7, this time with numbers. These numbers for a specific case can be obtained by actual measurements of a round trip pulse echo from a true point target, or they can be simulated. To obtain the first ellipse axis, the full width at half maximum (FWHM) of the pulse envelope along z provides a metric for axial resolution and gives

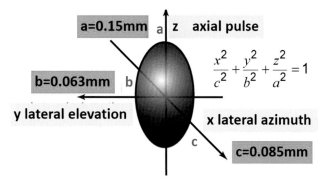

Figure 8.7 Numerical ellipsoidal realization of a point spread function at a focal point with a delineation of major elliptical axes Szabo (2014h). *Adapted from Szabo, T. L. (2014h). In* Diagnostic ultrasound imaging: Inside out *(2nd ed.). Elsevier.*

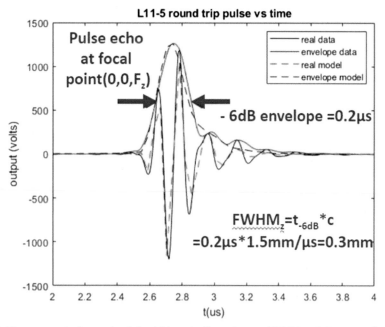

Figure 8.8 The arrows indicate the full width at half maximum (FWHM=2a) from a pulse envelope along the z ellipsoidal axis. It is also the measured value of the temporal width multiplied by the speed of sound, c, in the figure.

the value $2a$. Note in this case the simulation and experiment are nearly identical at the -6 dB or half amplitude points as shown in Fig. 8.8.

Beamplots can be used to obtain the other axes of the ellipse. The round trip FWHM along x can be obtained from the beamplot measured in water in Fig. 8.9 for

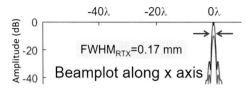

Figure 8.9 Full width at half maximum for round trip beamplot along the x "c" axis.

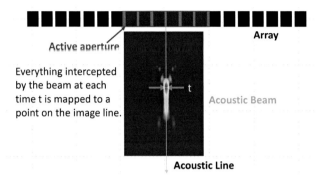

Figure 8.10 Formation of a single acoustic line in a scan. The active aperture shaded in red produces a three dimensional beam, here displayed as a two-dimensional section in the xz plane. The results of delay and sum processing for locations along the gray line are displayed as pixel intensities along that line. Results depend on the shape of the transmit beam and on the geometry of the targets.

$f = 8$ MHz, $L_x = 9.7$, $L_y = 13.1$, $F = 10$ mm. Similarly, if the azimuth and elevation foci are coincident, then the round trip FWHM in the yz plane is

$$\begin{aligned}
\text{FWHM}_{\text{RTY}} &= 0.88\lambda F/L_y \\
&= 0.88 \times 0.188 \times 10.0/13.1 = 0.126 \text{ mm.}
\end{aligned} \tag{8.2}$$

8.3.2 Formation of an acoustic line

What is represented by a pulse echo? Consider an active transmit aperture which is used to generate an acoustic line below it, drawn as a light gray line in Fig. 8.10. The size of the beam is mapped as an intensity plot shown in the azimuth plane xz. At a specific depth z a pulse hits a target and the echo returns at time $t = 2z/c_0$. What is shown in the image as the gray line is a one-dimensional (1D) time trace and not the beam (Fig. 8.10). This representation of the beam is only part of the story because of the 3D nature of the beam. Consider a thin disk reflector placed perpendicular to the z axis at distance z. The shape of the transverse beam in the xy plane at this depth is displayed at $z = 87\lambda$ in Fig. 8.11, the position highlighted in the previous figure at time t. Five 3D disk targets of different diameters are shown, and the thought

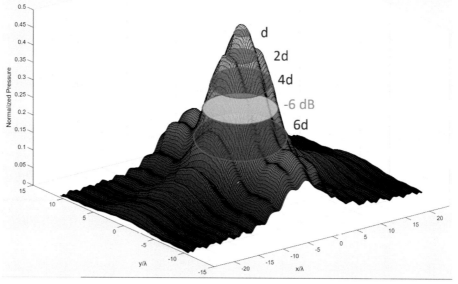

Figure 8.11 Transverse beamplot in *xy* plane at a depth of 87λ. Target disks of different diameters are superimposed on the two-dimensional beamplot.

experiment has each of the targets placed at depth z, one at a time, and poses the question "how much power is reflected back, as a function of target size?"

This figure suggests that the structure of the 3D beam and the size of the target may be important as to what constitutes the amplitude of the echo. Previously in Section 2.4.1, the reflection factor RF, dependent on impedance differences at a boundary was introduced; however, this was a 1D viewpoint. Put another way, this simplistic situation represents an infinitely wide target in the x and y dimensions, capturing all the transmitted power and reflecting it back as RF to be included in the A-line corresponding to depth z.

The targets shown in Fig. 8.11 intercept parts of the beam, but less and less as their diameter shrinks. What this figure shows is that as the diameter of the target changes, the part of the beam it intercepts also changes. Only that part of the beam defined by the size of the target is reflected back. Examination of the different targets reveals a different field distribution for each one; therefore, the strength of the reflection is proportional to not only the area of the target but also the shape of the field incident on it. The target can be thought of as an aperture reradiating toward the receiving transducer with an apodization shaped by the incident beam. The beamplot PSF axes $2b$

and $2c$ would form an elliptically shaped target in the beam (Fig. 8.9) at the (-6 dB) half-amplitude pressure level of the beam. A reflector of a size corresponding to the -6 dB beamwidth is highlighted in green in Fig. 8.11. In the limit, a tiny target would only sample the value of the peak pressure at the center of the beam. The location or size of the target (and its reflectivity) can be located anywhere in the beam, not just its center; however, whatever the scattering is, it will still be mapped to a pulse echo value at the roundtrip time t on an image line as shown in Fig. 8.10. Finally, consider the case where there is no scatterer anywhere in the main beam but if there is a strong one in a grating lobe such as those in Fig. 7.11 or 7.13, the echo will still be mapped into time t on the image line as though it were an echo from a target on the A-line.

The strength of reflection carried by the pulse echo depends on where the scattering object is in depth and is strongest in the focal region. Even if a second transducer having the same size as the original transmitter is close to it, as illustrated in Fig. 8.12, it cannot recapture all of the original launched energy; a small amount is lost at the edges due to diffraction. If the separation of the receiving transducer is increased, the

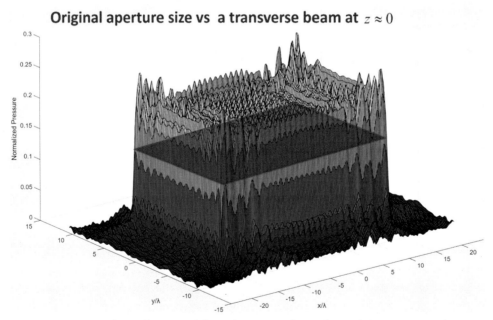

Figure 8.12 Acoustic beam from a 32λ by 20λ aperture focused at 64λ in transverse plane near $z = 0$. Shown in red is a rectangle with the original aperture size. Note that the field extends beyond the aperture shape.

result can be calculated as "diffraction loss" (Szabo, 2014c). If a pulse echo transducer radiates to a mirror or reflecting surface the same size as its aperture, the diffraction loss is computed at a distance $2z$ and is further multiplied by the reflection coefficient between the mirror material and the propagation medium.

8.4 Imaging point targets

8.4.1 Wire targets

In order to simulate point targets in a practical phantom, tiny filaments or wires are viewed in cross-section in an image plane. As explained in Section 2.3.1, a cylinder viewed in cross-section is a circle, and if small enough, it appears like a point as in the left side of Fig. 8.13. Because of the PSF, the tiny target is not reproduced exactly as a point in the image but appears as a distorted version, one of which is highlighted in the region of interest (R.O.I.) in the image on the right side of Fig. 8.13.

A magnified view of the R.O.I. has approximate image lines superimposed in Fig. 8.14. By taking a horizontal cut through this ultrasound image of a phantom wire target, the data in the image are sampled and a miniature beamplot results (here, the sparsely sampled beamplot is interpolated to smooth the plot of pixel amplitude vs horizontal pixel number). In this case, the image has a linear not a logarithmic amplitude scale selected (no compression).

Usually, a beamplot is obtained by scanning a sensor or point target through a beam; however, here a different but equivalent approach is used. A point target is located (here in the center of the image) in a fixed location. Because the same-shaped

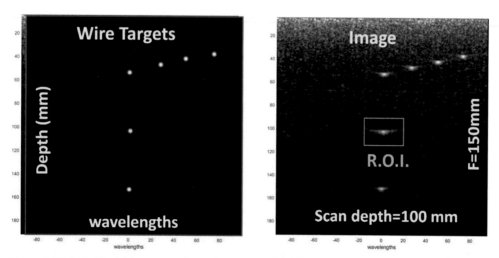

Figure 8.13 Left: Ideal wire targets in a phantom. Right: Ultrasound image of wires in a phantom with region of interest (R.O.I.) highlighted at depth of 100 mm.

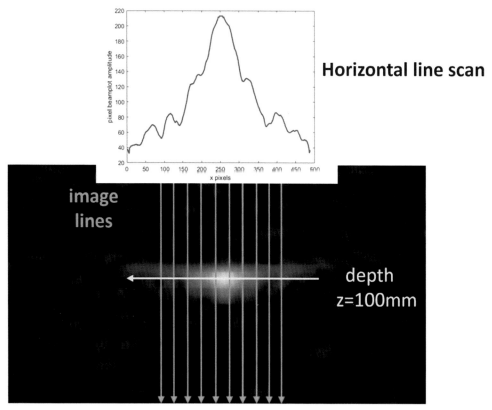

Figure 8.14 Miniature beamplot from R.O.I. of Fig. 8.13 created by sampling along a horizontal line drawn through image target as pixel amplitude vs x. Green vertical lines simulate the approximate location of the image lines through the target. Scan conversion interpolation fills in the values for pixels between lines.

beam appears in each acoustic line sequentially, it seems to be scanning past the wire target as illustrated by the top graphs in Fig. 8.15. This approach is equivalent to the one depicted in the bottom of Fig. 8.15, in which the beam remains fixed, and the sensor/target slides through the beam. Once again, the beamplots have been interpolated for clarity.

8.4.2 Scatter Image Simulator

To simulate images of point targets, the Scatter Image Simulator was created, based on the approach described by Zhu et al. (2012) and modified to add absorption (Szabo et al., 2013). As in a real image, beamformed lines are launched vertically downward to the scan depth. The center frequency, number of elements, and their pitch (period) and apodization as well as the propagating medium can be selected. The transmitting

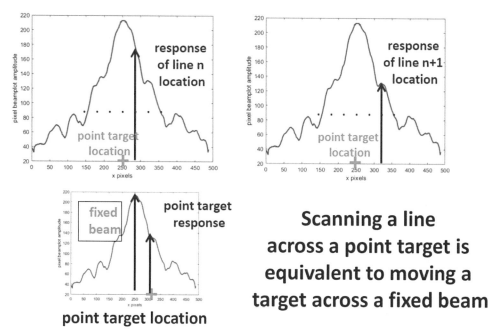

Figure 8.15 Two beamplot methods. Top: Scanned beam and fixed target. Bottom: Scanned target and fixed beam.

and receiving elements, and focusing arrangements are assumed identical. Point targets spaced at 1 cm intervals in depth are the default configuration; however, a free-wheeling option allows the user to create more targets arranged anywhere within the image boundaries. Finally, image controls for compression and contrast are available to permit examination of the PSF in detail and over a wide dynamic range. In the example shown in Fig. 8.16 the target at 2.5 cm is closest to the selected focal length of 3 cm. This simulator offers many variables for exploration, though a few are determined as default values to simplify the number of options. The image simulation is a function of both the array parameters, mentioned for example in Section 7.5.2, as well as compression and contrast image variables.

8.4.3 MultiFocus Simulator

In traditional delay and sum imaging using a set of scanlines, there is only one transmit focal length and this produces high-quality images over a relatively small depth of field near the focal depth. To improve on this situation, one approach is to split the image into horizontal strips or zones into each of which a transmit focal point is placed. Finally, all the zones are combined into one image. In other words, the highest resolution part of the image from each transmission is captured in each zone and all of them

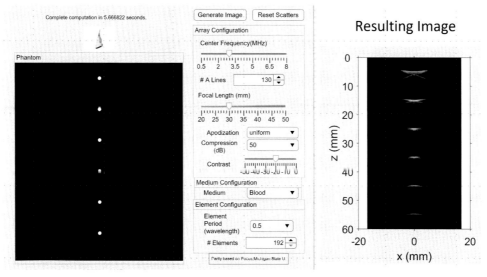

Figure 8.16 Left: Scattter Image Simulator graphical user interface showing arrangements of targets. Right: Simulated ultrasound image of the selected target arrangement.

are assembled into an overall image. The image looks better but it is obtained at the expense of frame rate. If the scan depth for each transmission is equal to the focal length and an additional distance to capture some of the depth of field of the focal zone, $F_n + \Delta z_n$, then the overall N multizone frame rate is slowed to

$$\text{FR}_{MZ} = \left[\sum_N 2(F_n + \Delta z_n)/c_0 \right]^{-1}. \tag{8.3}$$

Once again, for the purposes of the simulator (Zhu et al., 2013), the transmit and receive focusing are the same; however, there is in reality, no time penalty for focusing on receive and that part of focusing simulates dynamic receive focusing. Typically, in order to keep resolution constant with depth, a constant $F\#$ on receive is kept in each zone.

The options for this simulator are a superset of the previous simulator in that all the variables of that simulator are included as well as a new one, beamformer speed of sound. Recall that usually for diagnostic ultrasound, beamformers use a standard average speed of sound, $c_0 = 1.54$ mm/μs. However, in some cases, the speed of sound is considerably different; for example, in fat it is 1.43 mm/μs; the beamformer sound speed can be tuned or adjusted to improve the focus. The MultiFocus Simulator compares the multifocus approach to that of a single focus as evident in Fig. 8.17. Furthermore, the details of the response at a single selected scatterer for the multifocus approach can be examined in detail for both lossy and lossless media as illustrated in Fig. 8.18.

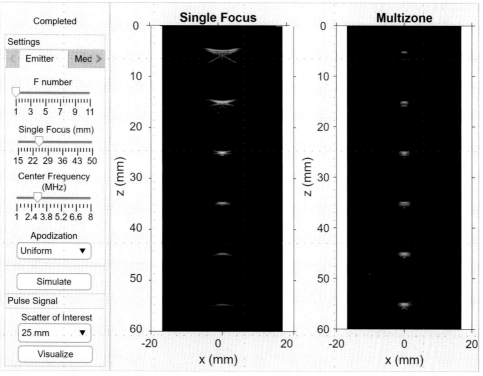

Figure 8.17 Multifocus Simulator showing the transmit panel and the display images for the settings shown on the right and a single focus image on the left.

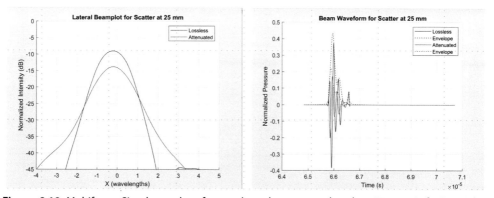

Figure 8.18 Multifocus Simulator plots for a selected scatterer placed at 25 mm. Left: Beamplot. Right: Axial pulse and envelope. Blue curves represents an unattenuated medium and red represents blood.

8.5 Time gain compensation

8.5.1 Time gain compensation amplifiers

A major challenge of imaging tissues with inherent absorption is the presentation of weak, attenuated pulse echoes at deeper depths. Compressing image data into a logarithmic scale helps but is not enough. Fortunately, power law absorption, Section 3.5.1, plotted on a logarithmic scale, also falls off linearly with depth (Szabo, 2014d). The solution most commonly implemented is called time gain compensation (TGC) (Szabo, 2014b). This approach begins by dividing the image into several (here eight) horizontal strips as shown in Fig. 8.19. An amplifier is assigned to each strip and is controlled by an individually adjustable amplifier gain slider. The individual sliders are adjusted to increase gain at each depth to produce a uniform level of background grayness throughout the image, and therefore, the formerly attenuated regions are recovered and visible.

Fig. 8.20A presents a graph showing the increasing attenuation loss with distance (Szabo, 2014b), as well as the TGC gain controls adjusted to compensate for it in a staircase pattern. Fig. 8.20B presents overall combined attenuation and TGC curve. In many ultrasound systems, the eight amplifiers are replaced by one voltage-controlled variable gain amplifier (VGA), and the control voltage is provided by a digital to analog converter that is programmed using the TGC slider values acting as control points.

Essentially, a smoothed voltage waveform is provided to the VGA as a function of time to modify the gain as the echoes return. The VGA is usually "logarithmic," that is, a linear change in the control voltage produces an exponential change in gain.

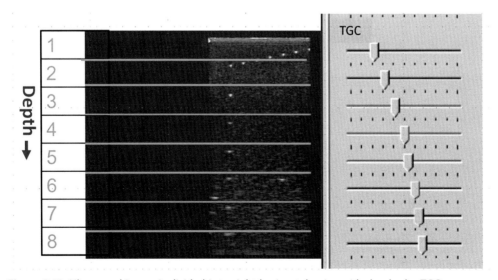

Figure 8.19 Ultrasound image is divided into eight horizontal strips with depth, the TGC zones.

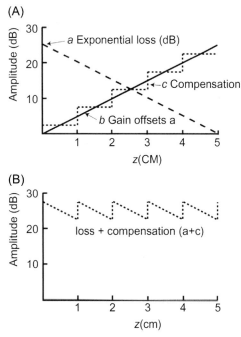

Figure 8.20 (A): Curve a is the exponentially decaying absorption with depth presented on a logarithmic scale. Curve b shows the average TGC gain offsetting the absorption loss. Curve c shows the staircase gain per strip vs depth. (B): Resultant loss and compensation curves (a + c) with depth. *From Szabo, T. L. (2014c). Diagnostic ultrasound imaging: Inside out (2nd ed.). Elsevier.*

8.5.2 Attenuation compensation

As explained in Section 3.5.1, absorption not only reduces the amplitudes of pulses propagating in a lossy medium but also changes their shape. A secondary effect, as shown in Section 6.5, is that absorption affects the beam and focusing. TGC amplifies pulses within each strip but it does not compensate for pulse and beam distortion caused by absorption and diffraction.

8.5.3 Time gain compensation simulator

The TGC Simulator depicted in Fig. 8.21 shows a restricted control panel which provides limited controls to rapidly generate an image with a single identical transmit and receive focus. Input options include a selection of images which correspond to different sets of beamformed data. On the right are the TGC gain sliders and the compression and contrast controls.

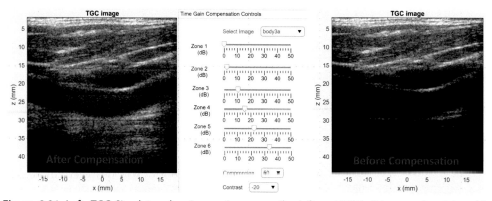

Figure 8.21 Left: TGC Simulator showing an image on the left and TGC sliders on the right with image adjustments lower right and image selection on the upper right. Right: Original image without TGC.

8.6 Scattering

8.6.1 Types of scattering

The similarity of the acoustic properties of tissues is primarily due to their high water content. The main constituents of the body are water (60%), protein (17%), and lipids (15%) (Greenleaf & Sehgal, 1992). In addition to blood, cells are bathed in fluid (interstitial) and have fluid within them (intracellular), as well as minerals and ions. Groups of similar cells (the basic building blocks) are organized into tissues. Different types of tissues are combined to perform specific functions as an organ, such as the heart or liver.

This commonality allows ultrasound to penetrate deeply into layers of tissue and offer back weak pulse echo reflections which can be observed and imaged with high gain receivers and amplifiers. The image itself is an interplay between the size of the tissue structures examined relative to an ultrasound wavelength (the approximate limit of resolution). Here we use the wavenumber k and the size of the scatterer of radius a (in wavelengths) to classify the scattering regime through the dimensionless product $ka = 2\pi a/\lambda$.

Greenleaf and Sehgal (1992) proposed a useful classification scheme for ultrasound tissue scattering. In their terminology, Class 0 scattering is associated with molecular solvent effects on a length scale of 10^{-4} Å (one Angstrom is 10^{-10} m). This type of scattering is due to macromolecular effects, which produce absorption and sound speed dispersion (discussed in Chapter 3). While we cannot see this class of scatterers directly, its presence is betrayed indirectly either by absorption with depth or by the need for more amplification with increasing depth. Class 1 scattering is caused by the concentrations of living cells being higher than 25 per resolution cell, and it is diffusive according to its length scale, $ka \ll 1$. Class 2, for scatterers on the scale of $ka \approx 1$, is scattering from the structure of tissue in concentrations on the order of one per resolution cell, and lies in an intermediate range between the Class 1 and 3 extremes. While Class 1 scatterers would result in speckle (unresolved scatterers discussed in the next section) or measurable aggregate (combined) effects, Class 2 scatterers are independent and distinguishable through their unique space- and frequency-dependent characteristics in an image. Class 3 scattering is specular on a length scale $ka \gg 1$, and is associated with organ and vessel boundaries such as the profile of the fetal face and what we would associate with ray propagation. A fifth category, Class 4, applies to tissue in motion, such as blood.

In a typical ultrasound image such as the one of fetal breathing shown in Fig. 8.22, can be found examples of the four scattering classes (Szabo, 2014e). From the necessity to compensate for absorption, as indicated by the use of TGC controls (not shown but implied and just discussed in Section 8.5), we conclude that Class 0 scatterers are present. The background speckle pattern (explained in the next section) indicates unresolvable Class 1 scatterers ($ka \ll 1$). Resolvable fine features such as the small blobs, vessels, and point targets shown in Fig. 8.22 correspond to resolvable Class 2 scatterers. Finally, the fetal facial profile is an example of a Class 3 scatterer. The dash of blue in the image is an indication of the inward flow (about -10 m/s) of amniotic fluid (fetal breathing) revealed by Doppler flow imaging and colors overlaid on the image, Szabo (2014f). Doppler and this type of imaging will be explained in Chapter 9.

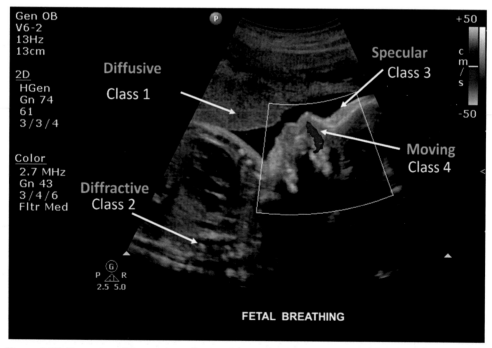

Figure 8.22 Examples of four types of scattering in the image of a fetus breathing amniotic fluid. *Courtesy Koninklijke Philips N.V.*

Terms that are used frequently for representing the appearance of tissue organization are "homogeneous, inhomogeneous, and heterogeneous" and boundary (Szabo, 2014f). A boundary is the border between two different or dissimilar materials. Waves crossing region boundaries may experience reflection and transmission effects, possible mode conversion, refraction, and changes in sound speed and absorption, according to the appropriate length scales. Tissue (or material) regions that have the same values everywhere or can be represented by one value of a parameter at every spatial point are "homogeneous." Global values of parameters are assigned to each region that is homogeneous. So far, most of the materials used for examples we have discussed have been homogeneous. The term "inhomogeneous" is used for tissues that are predominantly the same type with small fluctuations about a mean value. For example, in Fig. 8.22, the region above and halfway down the arrow pointing to Class 2, is inhomogeneous. A region enclosing a group of contiguous regions with different characteristics is called "heterogeneous." In this case, the tissue properties of the enclosed region vary with spatial position either through smaller subregions or, in the limit, from point to point. In Fig. 8.22, the region above the tip of the arrow pointing to Class 2, is heterogeneous.

8.6.2 Speckle

Speckle is a Class 1 scatterer which means that an image of an individual particle cannot be resolved but the effect of the particle can be seen (Szabo, 2014g). Speckle appears as the mottled background texture as shown in Figs. 8.14 and 8.19. In these cases, a phantom with wire targets is shown. This phantom is made from an initially homogeneous liquid material into which small subresolution micron size scatterers are randomly mixed and then it is hardened into a solid or gel. While these scatterers cannot be individually resolved, they have a cumulative group aggregate effect that can be imaged. Because the particles do not move, their effect is deterministic. A transducer returning to the same speckle region under the same settings will produce exactly the same speckle image. While viewers may believe they are seeing a characteristic texture of tissue or of a phantom, in fact, they see a pattern mainly determined by the geometry and bandwidth of the transducer and the focusing pattern (if any) of the beam.

The density of the number of particles in a region of a phantom can be altered to achieve different effects. A nonreflecting, scatter-free region will be speckle-free and appear dark. A region with a higher density of particles will appear more reflective or brighter than its surrounding material which has an average density of scatterers. Differences in such backscatter strength are referred to as having "contrast," and contrast resolution is the ability of an imaging system to display and distinguish small differences in contrast as discussed in Section 8.7. Most diagnostic imaging test phantoms have regions with differing Class 1 particle densities to provide a series of contrast targets to be compared to the background value.

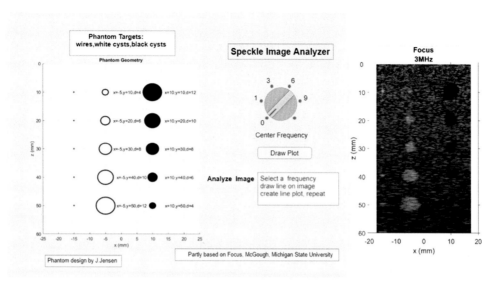

Figure 8.23 Left: Speckle Simulator GUI. Target pattern, frequency switch and line plot option. Right: 3 MHz resulting phantom image simulation.

8.6.3 Speckle Simulator

The Speckle Simulator provides examples of three types of targets against a speckle background. An example image is shown along with the graphical user interface (GUI) of the simulator in Fig. 8.23. The same kind of algorithm used in computing the Scatter Image Simulator is applied here. Jensen and Svendsen (1992) developed a type of pointillism for ultrasound by which any type of object for imaging could be represented by a 3D array of true point targets (not wires) each individually weighted in amplitude as appropriate (Szabo, 2014e). For example, to represent a disk such as those shown in Fig. 8.11, a set of closely spaced subresolution point targets with the same amplitude set in a circular pattern could be used. More realistic representations of heterogeneous tissue involve some degree of randomness in positioning. For representation of speckle, random positioning and amplitude assignment and sufficient density are necessary, at the expense of increased computation time with increasing numbers of particles.

8.7 Image contrast

8.7.1 Contrast resolution

As an ideal pulse for imaging is short with no trailing time sidelobes as explained in Section 4.2.9, an ideal beam is narrow with no sidelobes. A beam with very low sidelobes gives a clear separation between regions with high contrast, near a light–dark boundary. In Fig. 8.15, the concept of a scanned beam was introduced. If the boundary is represented by a step function, one that rises steeply from one level to another, then a beam convolved with the boundary is the representation of that boundary in the image. Similar to how sidelobes add "wings" to the image of a wire target, as illustrated in Fig. 8.14, sidelobes of a beam prevent the image of a boundary from going full white or black (or the two actual levels on either side of the boundary) because of a spillover effect. The two levels in an image are linked through a reflection coefficient with impedances for each level.

"Contrast resolution," a term coined by Maslak (1985) is a measure of the ability of a beam to resolve objects that have different reflection coefficients and is typically taken to be the −40 dB (or −50 dB) round-trip beamwidth. Pulse echo imaging is dependent on the backscattering properties of tissue. To first order, the possibility of distinguishing different tissues in an image is related to the reflection coefficients of tissues relative to each other (such as those shown in Fig. 1.3). These often subtle differences occur at the −40 to −50 dB level. Consider three scatterers at reflectivity levels of 0, −20, and −40 dB. If the main beam is clear of sidelobes down to the −50 dB level, then these three scatterers can be cleanly distinguished. If, however, the beam has high sidelobes at the −13-dB level, then both weak scatterers would be lost in the sidelobes. The level of the sidelobes sets a range between the strongest scatterers and the weakest ones discernible. In other words, the sidelobe level sets an acoustic clutter floor in the image.

Sidelobes appear as "wing-like" artifacts on either side of the main lobe such as those in the image of Fig. 8.14. Fig. 8.24 shows that the sidelobes can be reduced by apodization at the cost of a wider main lobe.

Examples of round trip beamplots from a 3 MHz broadband pulsed array are shown in Fig. 8.25. Recall the −6 dB beamwidths, here for the round trip case are given by

$$\text{FWHM} = b_{\text{rt}} \lambda z / L, \tag{8.4}$$

where $b_{\text{rt}} = 0.886$ for the unapodized case and $b_{\text{rt}} = 1.646$ for the Hamming case. For the same resolution, the Hamming aperture would need to be nearly twice that of the unapodized one. Here the −50 dB beamwidth for the unapodized case is about 63% wider than the apodized case which is far superior for contrast detail.

8.7.2 Contrast measurement

Beams with higher sidelobes "muddy the waters." In other words, there is a spillover that occurs in adjacent regions of high contrast due to the contrast resolution effect. For example, in the phantom image of Fig. 8.23 created by the Speckle Simulator, a horizontal line can be drawn through the 3 cm deep objects. The wire target appears at the left against a speckle background. Next is the "white cyst" with a line indicating its physical width and location. To the right, is the "black cyst" or black hole which, because of an absence of scatterers, should ideally register no echoes. The simplest definition of image contrast is to compare the mean image level in the object to that of its surroundings (Szabo, 2014g).

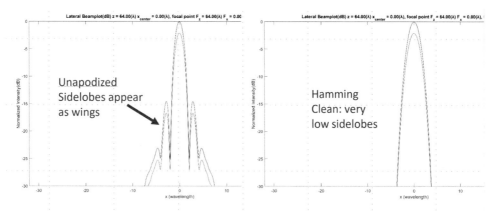

Figure 8.24 Comparison of 5 MHz, $L = 32\lambda$, $F\# = 2$ roundtrip pulsed beamplots for (left) unapodized aperture and (right) Hamming apodization.

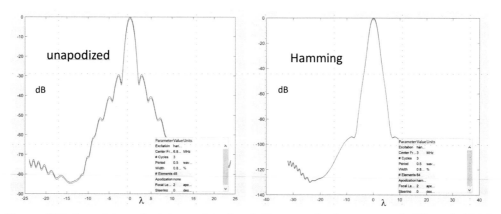

Figure 8.25 Roundtrip pulsed beamplots for (left) unapodized aperture of 17λ and (right) Hamming apodization of 32λ. Note that dB scales are different between plots, going to -90 dB on the left, and to -140 dB on the right.

The contrast ratio (CR) serves this purpose: compare the average gray-scale brightness level in the object, $\hat{A}_{in}$, to the mean level of its surrounding, $\hat{A}_{out}$:

$$\mathrm{CR} = \frac{\hat{A}_{out} - \hat{A}_{in}}{\hat{A}_{out} + \hat{A}_{in}}. \tag{8.5}$$

Note that a value of ± 1 is for good contrast; values close to zero indicate poor contrast. The gray scale values have values from 0 to 255, as shown by the text annotation for amax and amin in Fig. 8.26. By drawing a line the length of the original

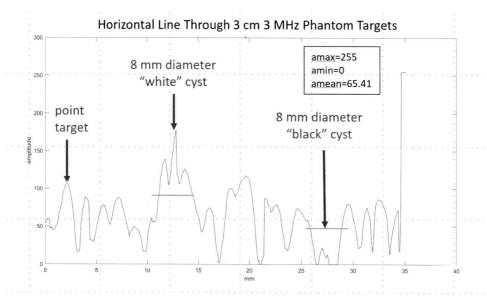

Figure 8.26 A horizontal line scan through a 3 MHz image of a phantom like that in Fig. 8.23, at a depth of 3 cm. Three objects are identified: a wire (point) target, an 8 mm diameter white cyst, and an 8 mm diameter black cyst. Also shown as green lines are the physical extents of the original cysts.

cyst diameter (shown by a green line in the figure), one can obtain the mean gray scale value across it. In this way, values for Eq. (8.5) can be obtained. Typical numbers for the white cyst are CR = −0.32 and for the black cyst, 0.20. Note other issues at play here, such as renormalizing the image to raise the highest gray scale value in the image to 255 (as indicated presently by the white value framing the picture at the right edge of the scan). Typical image settings of brightness and contrast will greatly influence the values of the CR metric and must be adjusted deliberately to make meaningful comparisons between images.

8.8 Ultrasound video

8.8.1 Delay and sum approach

As described in Section 8.2.2, the video framerate from a scanning acquisition and delay and sum beamformer consists of N lines shot sequentially per frame as given by Eq. (8.1). However, a standard video rate for a monitor is usually 30 frames per second (fps), so the ultrasound refresh rate may be different, depending on the scan depth. In conventional systems, much of the processing is done by specialized chips in parallel to achieve real-time operation. To give an estimate of the number of operations needed for a single frame, consider a 40 MHz digitization rate. For a 20 cm scan depth, an array of 128 elements and 128 image lines in a frame, this leads to about 109 million samples of information to be processed. Then at 30 fps, there are about 330 million samples/s and this does not include additional processing and filtering.

A second problem alluded to in Fig. 8.2 is that a standard video frame, say for example, the National Television Standard Committee (NTSC) format contains a rectangular matrix of $m \times n = 720 \times 486$ pixels. Ultrasound data must be converted to a standard video format through a process called scan conversion. Also, for this example, of the 8000 samples in a beamformed line, only 486 at most are converted into pixels for this example, so the process is inefficient. In a research ultrasound system, real-time software beamforming is a challenge, and parallelization of the code or computation (parallel processors) is required. Also needed are suitable approximations that greatly reduce the number of operations performed with a conventional hardware-based beamformer. Computing delay and sum results over only the pixels needed to form the image is one such optimization, along with other approaches to accelerate performance as described in the Verasonics pixel-oriented processing patent (Daigle, 2017), and in detail in 10.2.2.

8.8.2 Plane wave compounding approach

In the plane wave compounding method described in Section 6.6.1, fewer transmissions are needed for coverage, corresponding to 7 (typically) to 31 lines. On the lower end, this is a speedup of about 18 times in frame rate or much faster than standard video. On receive, the plane wave pulse echo data frames are stored in memory, and then dynamic focus beamforming is applied to each frame prior to coherent summation and calculation of the amplitude to form the compounded image. In these, high speed capture of heart valve motion or details of blood flow can be played in slow motion to reveal diagnostically useful information. In order to obtain focusing resolution comparable to that of focused line imaging, more transmissions are needed, so in practice a compromise is struck between desired frame rate and resolution (Montaldo et al., 2009).

8.8.3 Ultrasound Video Simulator

Examples of ultrasound videos are presented by the Ultrasound Video Simulator. Computations are the same as those used for the Scatter Image Simulator except that motion is involved. Several ultrasound videos are provided with variable frame rates and a video player in this simulator as shown in Fig. 8.27.

8.8.4 Bewildering ultrasound video artifacts

Images of unexpected objects appearing in an ultrasound image can often be explained by basic physics. For example, if there are two objects at 4 and 8 cm below the transducer array as shown in Fig. 8.28, a single transmission will create unique pulse echoes from each one. If, however, the scan depth is repeated at 6 cm intervals (round trip

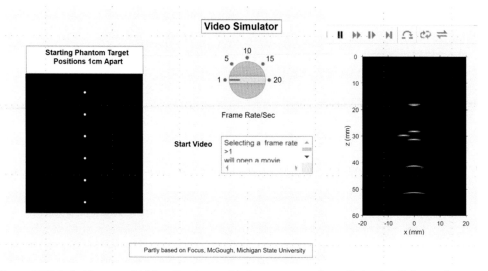

Figure 8.27 Left: Ultrasound Video Simulator with frame rate options. Right: A still frame from an ultrasound video.

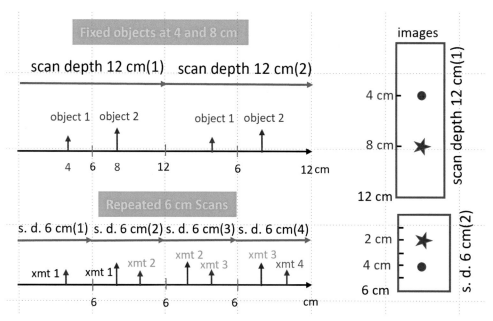

Figure 8.28 Top: Fixed locations of two objects and image with 12 cm scan depth. Bottom: Pulse echoes from different transmissions repeated at intervals corresponding to a 6 cm scan depth with image to the right.

time of 40 µs), then the second echo from the first transmission will appear in the second and subsequent images as if it was located at a depth of 2 cm. Using a deeper scan depth can eliminate this ghost object.

A second common artifact is reverberation, a concept already presented in Section 2.5.1. Pulses can undergo multiple internal reflections when trapped in a layer or chamber. When this phenomena occurs in an image, it adds a tail-like artifact which obscures whatever is beneath the layer in depth. Another form of reverberation happens when a low absorption region, for example, a fluid-filled volume or globule, is surrounded by a more absorbing medium that is used to set the TGC levels. The sound passing through and resonating in the globule is higher in amplitude so that its multiple echoes can form a white comet-like tail on the bottom of the globule. An example can be seen below the cyst in the upper right side of Fig. 8.6.

A third artifact has been discussed in Sections 7.4 and 8.3.2: grating lobes. These artifacts can present confounding lateral visual repetitions and distortions of objects being imaged, and the grating lobe structure moves with the transducer. Experiments on grating lobes are included in laboratory 6.

8.9 Lab 8: exploring ultrasound images and videos

8.9.1 Exercises with simulators

The **Scatter Image Simulator** calculates the images of wire "point" targets from arrays with selectable parameters; a fundamental step for simulating a practical imaging *system* response. Students have had experience with the Vantage system, changing transmit beam properties and examining the effect of the PSF on a wire target. This simulator provides a numerical tool to predict what a complete imaging system will produce. The students are encouraged to explore a range of combinations observed with the L11-5v transducer and compare with the images produced with the hardware. Exploration of imaging in lossy media with other transducers that operate at lower frequencies illustrate the tradeoffs between resolution and imaging depth due to the role of absorption.

For conventional line-mode scanning, we have seen that the transmit beam focal zone produces the best image quality as measured by the PSF and beamplots. To extend the favorable depth over which to obtain good images, stitching together results from multiple transmits with foci spanning a range of depths is done at the expense of frame rate. The **Multifocus Simulator** allows the student to visualize the improved resolution gained in this way throughout the overall image. They are guided in selection of parameters that help demonstrate the tradeoffs between frame rate and lateral resolution using multiple transmits.

As mentioned in Section 8.5.3, the **TGC Simulator** allows for exploration of the utility of TGC in a purely numerical environment. The students are already familiar with the controls: TGC sliders were provided in Lab 3 to permit them to optimize the images in order to be able to simply see what they were interested in, for example, reverberation in thin targets. Now, equipped with the basic understanding of the role and the implementation of TGC, they can explore overall image equalization in a simulation environment, to compensate for attenuation effects and for intensity emphasis due to beam focusing effects. Similarly, they are able to adjust the dynamic range of signals represented in an image using the "Compression" control. Image contrast can be separately adjusted by the "Contrast" control. Again, these manipulations will be compared with observations conducted using the Vantage.

Although speckle is a consequence of scattering from subresolution particles and filtering by the transducer's passband, it is far from merely "noise" in the image. In practice, scatterer size, distribution, scattering strength, and density all vary with tissue type and can provide texture differences across an image even though the perceived texture is not found in the material itself. Nevertheless, if all material properties are the same except for scatterer density, then one can evaluate the ability of an imaging system to distinguish different levels of backscatter within regions using the CR shown in Eq. (8.5). Algorithms that seek to delineate boundaries between regions of different speckle can be evaluated using contrast resolution phantoms for which the geometry of the boundaries is known, and using numerical simulations such as are provided through the **Speckle Simulator**. The students generate plots like those in Figs. 8.23 and 8.26 for a number of contrast and compression settings to gain an appreciation for the effectiveness of these controls in helping distinguish different textures.

Finally, the students can run through some of the options with the **Ultrasound Video Simulator** which presents movement in a simulation. Because ultrasound imaging produces images at a rapid frame rate, it is one of the few imaging modalities that can provide real-time visualization of the anatomy, including blood flow, as will be described in Chapter 9.

8.9.2 Experiments and exercises with the Vantage system

The laboratory module has once again extended its interface to include more imaging modes, and new image controls, as illustrated in Fig. 8.29. In addition to conventional scanned focused beams, the new modes include plane wave imaging using seven plane waves at different angles, compounded coherently or incoherently. Furthermore, nonlinear harmonic imaging, which is introduced in Section 10.3, is also offered. Conventional image optimization controls are also added, included digital gain (image brightness) and dynamic range compression for students to use in adjusting images and evaluating the contrast ratio of Eq. (8.5). Finally, a slider control for adjusting the sound speed used in the beamformer for image reconstruction is also added, where before the students had only the selection between water (1480 m/s) and phantom (1465 m/s) values.

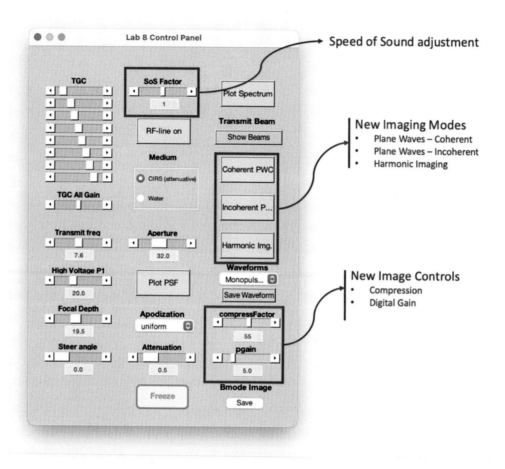

Figure 8.29 Screenshot of the Vantage GUI control panel for Lab 8. New ultrasound modes and new image controls are added. The beamformer sound speed can also be continuously adjusted.

One of the experiments uses the new sound speed adjustment to observe the effect of using an incorrect value for reconstruction. In Fig. 8.30, two images of the same view are created using different sound speeds, the one on the left at 1540 m/s, a value that is typically used for tissue imaging, and the image on the right using the value that corresponds to the rubber material sound speed for the phantom. One can clearly see the distortion of the circular shape of the dark cyst target and the characteristic blurring of the point target responses in the left image, and conversely, the image on the right looks properly scaled and focused. Indeed, the true separation of the wire targets is now rendered, and the students will be tasked to determine that distance as part of the procedure given their determination of the optimal medium sound speed and the travel times measured using the RF data line plot they have been using in prior labs.

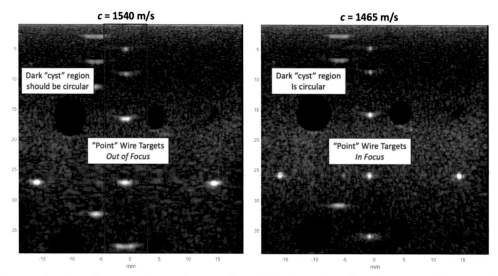

Figure 8.30 Left: Reconstruction with a sound speed that is different from the true medium sound speed results in an unfocused image, as is evident from the appearance of the point target PSFs, and the distorted cyst region. Right: Using the true sound speed for the phantom focuses the image and results in the best (smallest) PSF for the wire targets over all depths.

The TGC controls can also be used to compensate for attenuation, and for other mechanisms leading to brighter bands in the phantom images, such as the beam focal depth and the elevational focal depth. Fig. 8.31 illustrates the effect of attenuation on images when TGC is not used (flat TGC refers to setting all of the sliders to the same position, and thus no change in gain vs depth). In comparison, when TGC is adjusted to compensate for attenuation, the image brightness at depth is similar to that in shallower zones, as illustrated in Fig. 8.32.

We can also observe in Fig. 8.32 the differences in image appearance between focused line scanning and plane wave imaging. The focused lines result in a brighter band about the focal depth, where the image quality is also very good: the PSF is small, and the CR is high for the dark cyst targets. While the plane wave image produces PSF responses that are almost independent of depth, this comes at the expense of greater sidelobes that cause the cyst targets to show some clutter, with the associated reduction in CR. On the other hand, the frame rate for plane waves is, for the case of seven plane waves and 128 scanlines, a factor of $128/7 \approx 18$ times faster. This higher frame rate enables other kinds of investigations, for example, those in which motion in the medium is very fast.

Finally, Fig. 8.33 presents an illustration of the differences between coherent and incoherent compounding, using plane wave imaging as an example. Coherent compounding

Flat TGC

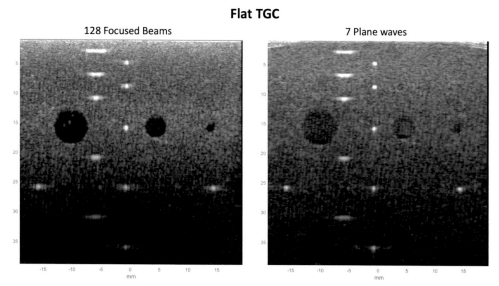

Figure 8.31 Without TGC to compensate for attenuation, images darken at depth. Left: Focused beam image. Right: Plane wave image.

Using TGC to compensate for Attenuation

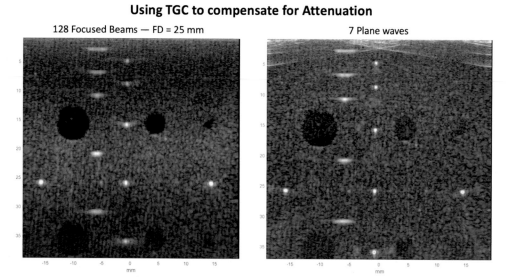

Figure 8.32 Using TGC to compensate for attenuation only (and not for the enhancement at the focal depth), images are brightness equalized over depth. Left: Focused beam image. Right: Plane wave image.

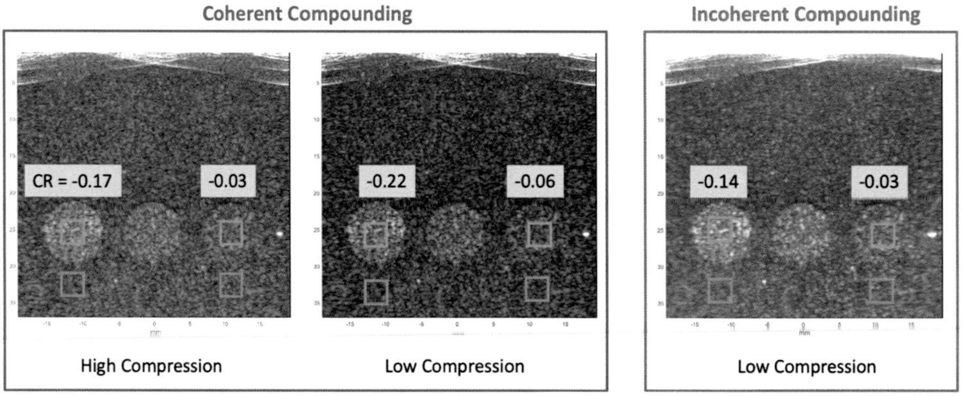

Figure 8.33 A comparison between coherent compounding of plane wave acquisitions (left, blue frame) and incoherent compounding (green frame). In the coherent case, two levels of dynamic range compression are used to illustrate the role of compression in imaging.

sums the results of each image reconstruction while preserving the phase of the beamformed RF, and this approach leads to brighter extended targets (scatterers of type 2, as described in Section 8.6.1) which appear similar at several plane wave angles and thus add coherently, but particles that produce speckle (type 1) will produce different responses and interfere with each other in the summation. This can produce high contrast images because the speckle tends to cancel and results in occasional nulls at some pixels in the image. Incoherent averaging sums the amplitudes only, thus fills in the nulls and averages the speckle at the expense of adding a hazy "floor" to the image brightness, diminishing contrast in favor of smoothing the speckle. Usually, systems combine coherent and incoherent compounding to find a compromise between speckle reduction and contrast preservation.

The students gain an appreciation for the different image characteristics by collecting images with each mode (coherent and incoherent) and then computing the CR for the same regions to quantify the differences. The CR values for the echogenic targets are labeled above the target, and the inner regions (inside the blue boxes) are compared to the surrounding regions (inside the orange boxes), as defined in Eq. (8.5). The view of the phantom is the same in each case, and we can easily observe that reducing compression increases contrast, and that coherent compounding yields higher contrast than incoherent compounding. In this exercise, students quantify the role of compression and gain adjustments on image contrast by making CR measurements for different settings. They will gain an appreciation for the challenges in comparing images, given all of the factors that determine the value of a given metric, in this case, the CR. Fig. 8.33 presents an example of CR values for different compression and gain settings, and for different acquisition modes, other parameters being equal.

References

Daigle, R. E., (2017, May 16). *Ultrasound imaging system with pixel oriented processing* (US patent 9,649,094 B2).

Greenleaf, J. F., & Sehgal, C. M. (1992). *Biologic system evaluation with ultrasound*. New York: Springer Verlag.

Jensen, J. A., & Svendsen, N. B. (1992). Calculation of pressure fields from arbitrarily shaped, apodized, and excited ultrasound transducers. *IEEE Transactions on Ultrasonics, Ferroelectrics, and Frequency Control, 39*, 262–267.

Maslak, S. M. (1985). Computed sonography. In R. C. Sanders, & M. C. Hill (Eds.), *Ultrasound annual 1985*. New York: Raven Press.

Montaldo, G., Tanter, M., Bercoff, J., Benech, N., & Fink, M. (2009). Coherent plane-wave compounding for very high frame rate ultrasonography and transient elastography. *IEEE Transactions on Ultrasonics, Ferroelectrics, and Frequency Control, 56*, 489–506.

Szabo, T. L. (2014a). *Chapter 2. Diagnostic ultrasound imaging: Inside out* (2nd ed.). Oxford, UK: Elsevier.

Szabo, T. L. (2014b). *Chapter 10. Diagnostic ultrasound imaging: Inside out* (2nd ed.). Oxford, UK: Elsevier.

Szabo, T. L. (2014c). *Chapter 4. Diagnostic ultrasound imaging: Inside out* (2nd ed.). Oxford, UK: Elsevier.

Szabo, T. L. (2014d). *Chapter 6. Diagnostic ultrasound imaging: Inside out* (2nd ed.). Oxford, UK: Elsevier.

Szabo, T. L. (2014e). *Chapter 9. Diagnostic ultrasound imaging: Inside out* (2nd ed.). Oxford, UK: Elsevier.

Szabo, T. L. (2014f). *Chapter 11. Diagnostic ultrasound imaging: Inside out* (2nd ed.). Oxford, UK: Elsevier.

Szabo, T. L. (2014g). *Chapter 8. Diagnostic ultrasound imaging: Inside out* (2nd ed.). Oxford, UK: Elsevier.

Szabo, T. L. (2014h). *Chapter 7. Diagnostic ultrasound imaging: Inside out* (2nd ed.). Oxford, UK: Elsevier.

Szabo, T. L., & Lewin, P. (2013). Ultrasound transducer selection in clinical imaging practice. *Journal of Ultrasound in Medicine: Official Journal of the American Institute of Ultrasound in Medicine, 32*, 573–582.

Szabo, T., Nariyoshi, P., & McGough, R. (2013, July 22). Acoustic beam simulator with aberration, power law absorption, and reverberation effects. In *IEEE ultrasonics symposium proceedings* (pp. 374–377). Prague, Czech Republic.

Zhu, Y., Szabo, T. L., & McGough, R. J. (2012, October 7–10). A comparison of ultrasound image simulations with FOCUS and Field II. In *2012 IEEE ultrasonics symposium proceedings* (pp. 1694–1697). Dresden, Germany.

Zhu, Y., Szabo, T. L., & McGough, R. J. (2013, July 24). Multiple zone beamforming in FOCUS. In *2013 IEEE ultrasonics symposium proceedings* (pp. 1464–1467). Prague, Czech Republic.

CHAPTER 9

Doppler

9.1 Overview

9.1.1 The block diagram revisited

In general, we use the term "Doppler" as a catchall word meaning flow measurement and imaging with ultrasound. Making Doppler images requires unique waveforms and signal processing. Often Doppler images and data are combined with or overlaid on conventional images and this involves a juggling act; processing interleaved image and Doppler signals. All of the block diagram shown in Fig. 9.1 is involved: from

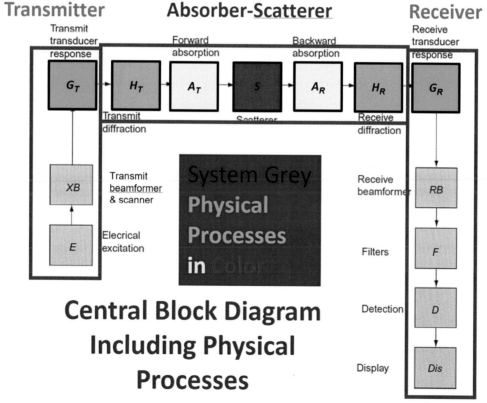

Figure 9.1 Central block diagram highlighting all blocks involved in Doppler processing (Szabo, 2014a). *From Szabo, T. L. (2014a). Overview. In Diagnostic Ultrasound Imaging: Inside Out (2nd ed., pp. 39—54). Elsevier.*

Essentials of Ultrasound Imaging
DOI: https://doi.org/10.1016/B978-0-323-95371-9.00007-2

generating unique Doppler waveforms, usually at a separate center frequency than that used in imaging in block E, all the way through the beamforming in parallel with imaging until display. At the filtering stage, Filter block F, dedicated signal processing extracts Doppler information from the data and prepares it for Doppler images and data presentation.

9.1.2 The Doppler effect

What if the source of the expanding spherical wavefront source introduced in Section 2.3 suddenly started moving to the right with a velocity of c_s? One would see the wavefronts change as shown in Fig. 9.2. In the mid-nineteenth century, Christian Doppler determined how the consequence of source motion was a perceived difference in the frequency depending on the location of the observer. This illusion applies to both acoustic and optical waves, though the physics of relativity makes the situation different in optics, and the apparent shift in frequency is now called the "Doppler effect".

Pierce (1989) has shown that the perceived frequency shift Δf is related to the vector dot product of the source ($\mathbf{c}_s$) and unit observer ($\mathbf{u}_0$) vectors, which differ by an

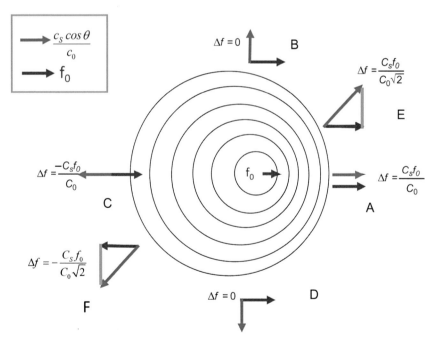

Figure 9.2 Wavefronts for a source initially emitting a frequency f_0 at rest. The perceived Doppler frequency shift is shown for several observer positions around the source moving at c_s. The propagation speed of sound for the medium is c_0. It is assumed that $c_s \ll c_0$. *Adapted from Szabo, T. L. (2014b). Section 11.2. In Diagnostic Ultrasound Imaging: Inside Out (2nd ed.). Elsevier.*

angle θ thus introducing a factor of $\cos\theta$. Pierce also solved for the situation in which moving fluid in a stationary medium is imaged using ultrasound.

To understand the expression for the Doppler shift Δf heuristically, we first recall the relationship between the wavelength λ_0, the ultrasound frequency f_0 (with period T_0) and the speed in the medium c_0:

$$\lambda_0 = T_0 c_0 = \frac{c_0}{f_0}. \tag{9.1}$$

Now we place ourselves alternatively in two frames of reference and look at the parameters of Eq. 9.1 in each of those contexts. First, in the frame of the moving source we see that, in the direction of motion (ahead) or $\theta=0$, the wavefronts appear to bunch up as the source velocity c_s increases. This is because the medium in which the waves propagate is stationary and cannot support waves moving faster than the speed of sound, and so the velocity of the wavefronts moving ahead of the source appears to slow down; in the limit, the wavefronts bunch up together into a shock front as the source speed approaches the speed of sound. The apparent velocity of propagation ahead of the source is $c_0 - c_s$, but riding along with the source, the emitted frequency f_0 is unchanged. Thus

$$\lambda_{\theta=0} = \frac{c_0 - c_s}{f_0}. \tag{9.2a}$$

When viewed by an observer on the ground, the stationary medium sound speed is c_0, but the wavelengths appear shorter, thus seemingly generated by a source of frequency $f_0 + \Delta f$, such that

$$\lambda_{\theta=0} = \frac{c_0}{f_0 + \Delta f}. \tag{9.2b}$$

Equating Eqs. 9.2a and 9.2b, we obtain

$$\frac{c_0}{f_0 + \Delta f} = \frac{c_0\left(1 - c_s/c_0\right)}{f_0} \tag{9.3}$$

and

$$f_0 + \Delta f = f_0\left(1 - c_s/c_0\right)^{-1} \cong f_0 + f_0 c_s/c_0, \tag{9.4}$$

where we require that $c_s \ll c_0$. Finally, we obtain the relation for the Doppler shift Δf in the forward direction, $\theta=0$,

$$\Delta f_{\theta=0} = f_0 \frac{c_s}{c_0}. \tag{9.5}$$

From A in Fig. 9.2, the frequency increases toward the right, leading to smaller times between wavefronts. In the opposite direction since the shift is negative, the frequency is lowered and the wave period increases. Above and below the source at positions B and D, there is no shift; the perceived frequency is f_0. The lettered locations around the source follow the expected vector dot product relationship as shown; more information can be found in Szabo (2014b). In general, for an observer at angle θ with respect to the direction of source motion, the Doppler frequency shift is

$$\Delta f(\theta) = f_0 \frac{c_s}{c_0} \cos\theta. \tag{9.6}$$

As an example, the perceived frequencies for the observers shown in Fig. 9.2 can be calculated for a 500 Hz source tone moving at a speed of 100 km/hour (v=27.78 m/s) in air (c_0=330 m/s). Observers B and D, at 90 degrees to the source vector, hear no Doppler shift. Observer A detects a frequency of 542.1 Hz, while observer C (here, $\theta=\pi$) hears 457.9 Hz. This shift may seem slight but is highly perceptible by human hearing, as it corresponds to a front (observer A) to back (observer C) difference of about three musical half-steps.

9.1.3 In This Chapter You Will Learn

An important diagnostic application of ultrasound is the real-time noninvasive measurement and visualization of flow. The Doppler effect, the apparent perceived shifting of a transmitted frequency dependent on the speed of the source and the position of the observer, is put to good use in pulse-echo ultrasound. Pulse wave Doppler has the advantage over continuous wave (CW) Doppler methodology by isolating the region of interest through the selectable direction and width of a time gate. Visualization approaches, Color flow imaging and power Doppler imaging, which enable the depiction of complicated flow patterns in vessels and the chambers of the heart in real-time, are introduced. How most of these methods (except CW Doppler) are constrained by geometry (angle of insonification) and sampling rate will be explained. Relatively new approaches such as Ultrafast Doppler which offers simultaneous capture of flow imaging and Doppler waveform data, and also Vector Doppler which displays local two-dimensional flow directions and magnitudes, are described. A Doppler Simulator provides interactive examples of the complicated interplay of flow geometry, insonification angle and sampling rate on Doppler imaging.

9.2 Principles of Doppler Ultrasound Measurement of Flow

Pulse-echo Doppler ultrasound may have been first mentioned by Constantin Chilowsky and Paul Langevin (2016) in their 1916 patent application for underwater

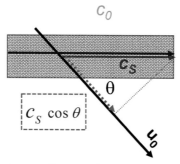

Figure 9.3 Sound propagating along a direction that intercepts blood flowing in a vessel at c_s at an angle θ. The projected speed (shown as a dashed line) along the direction of sound is $c_s \cos(\theta)$.

pulse-echo ranging (Duck, 2022). They mention a method to detect relative motion between the observer and target by comparing the Doppler-shifted frequency from the target to the frequency of a stable source.

To apply the Doppler principle to a pulse-echo configuration consider a thought experiment using Fig. 9.3. Imagine a stationary ultrasound transducer sitting at the upper left end of the black arrow which represents the direction of ultrasound transmission, at angle θ relative to the vessel, along a unit vector $\mathbf{u_0}$; blood, moving through the vessel at a speed of c_s, is crossed by sound moving at a speed of sound c_0. We assume the speed of sound in blood and the surrounding tissue are approximately the same, and the blood motion is shown as a vector $\mathbf{c_s}$. From the point of view of a tiny observer riding on a blood cell, the stationary transducer appears to be moving to the left at a speed c_s. The projected velocity of blood onto the sound direction is

$$v = \mathbf{c_s} \cdot \mathbf{u_0} = c_s \cos\theta \qquad (9.7)$$

This projection is seen as a dashed blue line in Fig. 9.3. The Doppler shift relationship is the same for a moving medium as it is for a moving source; the frequency of the transmitted wave appears shifted with respect to an observer moving with the blood, but in the case of Fig. 9.3 the blood is moving away from the sound source, so the shift is negative. In ultrasound pulse-echo imaging, the incoming sound is scattered back to the transducer, and now the source of the sound is tied to the blood which is again moving away from the transducer, and this contributes a second identical shift! Consequently, the Doppler frequency shift Δf_{RT} for the round-trip path in pulse-echo ultrasound is given by

$$\Delta f_{RT}(\theta) = 2f_0 \frac{c_s}{c_0} \cos\theta. \qquad (9.8)$$

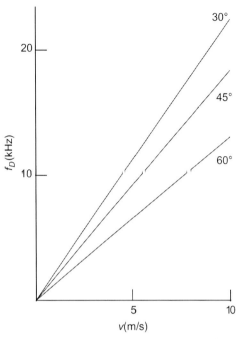

Figure 9.4 The Doppler shift frequency for a 2 MHz center carrier frequency ultrasound beam at different angles θ to a vessel with a blood flow velocity v. Here $f_D = \Delta f_{RT}$. Typical blood flow velocities are below 1 m/s, but reach much faster speeds in and near the heart. *From Szabo, T. L. (2014c). Section 11.4. In Diagnostic Ultrasound Imaging: Inside Out (2nd ed.). Elsevier.*

9.3 Continuous wave Doppler

Fortunately, Doppler shift frequencies for blood flow in human bodies fall into the audio range. Typical Doppler shift frequencies based on Eq. (9.8) for 2 MHz center frequency ultrasound can be seen in Fig. 9.4. Through signal processing and filtering, the shifted Doppler frequency is extracted. This is often done by introducing the transmitted frequency into a mixer with the received frequency to give a Doppler difference frequency. A Hilbert transform creates a quadrature signal which adds phase to amplitude information. Phase provides a way of including the directionality of flow.

While CW Doppler systems are sensitive, they have shortcomings. Because there is no time discrimination, anything in the acoustic beam path that moves contributes to the Doppler signal. This movement may include adjacent tissue motion. A popular CW Doppler transducer configuration is a pair of tilted separate "pencil beam" transducers whose acoustic paths, in the shape of an "X," intersect in an interaction region. The transducers are moved blindly (acoustic blindness because there is no time depth navigation) until the loudest Doppler signal is heard when the interaction region overlaps the vessel or desired flow region. Once located, the transducers and system can yield the measured fluid flow velocity and direction in real time as stereo audio output or in the form of a graphical Doppler velocity screen display. Find more information on CW Doppler in Szabo (2014c).

9.4 Pulsed wave Doppler and Doppler processing

9.4.1 Pulsed wave Doppler

Based on what we already know about ultrasound, Fig. 9.3 is too simplistic to capture the physics of all that is going on. First of all, there is a 3D ultrasound beam crossing the vessel not a line. Second, the vessel is full of red blood cells—shaped like double concave lenses about 7 μm in diameter and 2 μm in thickness and they have different attitudes and are constantly moving into other positions. Third, the blood flowing in the vessel is not uniform across the diameter of the vessel but has a parabolic-like velocity profile when flow is rapid, due to viscous drag from the stationary vessel wall.

The beginnings of a more realistic model are illustrated in Fig. 9.5. In the upper left of Fig. 9.5, a transducer (symbolized by a red line) emits a beam which traverses a vessel filled with flowing blood. The overlaid beam insonifies a volume of blood scatterers, and as in imaging, the total amount of scattering occurring at each pulse echo time t is mapped along the line now representing the acoustic axis of the beam. As in imaging, we now graduate from CW processing to add the time dimension. In pulsed

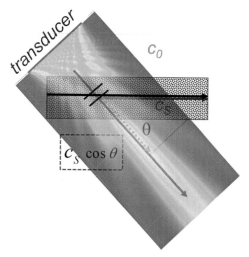

Figure 9.5 Ultrasound transmit beam intersecting a vessel with scatterers at an angle θ and showing the flow vector component projected along the sound direction as a dashed line. The two bars represent a Doppler time gate.

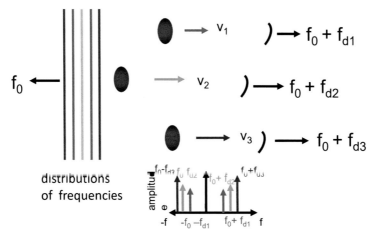

Figure 9.6 Red blood cells moving at different velocities cause a distribution of Doppler-shifted frequencies.

Doppler, tone bursts at a Doppler reference transmit frequency f_0 are repeated at selected regular intervals (represented by block E of the block diagram). Note in Fig. 9.4, the Doppler transmit frequency is 2 MHz (not to be confused with the Doppler shift frequency). Backscatter from these pulses can be observed from a selected region represented by a range gate (the parallel red lines in Fig. 9.5). The extent of the gate can also be adjusted using the length of the transmit burst to change the "sample volume" and listen in to blood flowing in a certain part of the vessel. Also, the direction of the Doppler line (and therefore the direction of the Doppler beam) can be changed, thus modifying the Doppler angle θ.

Cells moving at different velocities cause separate Doppler-shifted frequencies as depicted in Fig. 9.6. Because the Doppler beam and range gate encompass a volume of tumbling scatterers moving at different speeds, spectral broadening occurs.

9.4.2 Pulsed wave Doppler processing and display

The main difference between continuous and PWD is the advantage of localizing sampling in depth through temporal gating. Just as array elements behave as spatial samplers (as discussed in Chapter 6), finite length Doppler pulses act as time domain samplers. The repetitive nature of these pulses can be conveniently represented by the *shah* or sampling function in the time and frequency domains (Bracewell, 2014). This approach will also easily show the consequence of undersampling, which is the main limitation of pulsed Doppler.

As shown in Fig. 9.7A, pulses are repeated at intervals of T_{PRF} of $1/f_{PRF}$ where f_{PRF} is pulse repetition frequency and the time between pulses is the pulse repetition interval, $PRI = T_{PRF} = 1/f_{PRF}$. Each pulse has an adjustable length of M periods of

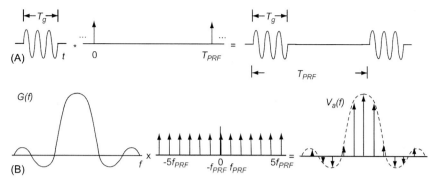

Figure 9.7 Pulsed wave Doppler transmit waveforms and spectra. (A) Repeating transmit pulse parameters. (B) Spectrum of repeating pulses. *From Szabo, T. L. (2014d). Section 11.5. In Diagnostic Ultrasound Imaging: Inside Out (2nd ed.). Elsevier.*

the transmit frequency, f_0, thus the gate length is

$$T_g = MT = M/f_0 \tag{9.9}$$

and typical lengths are longer than the one shown in Fig. 9.7A. Similar to the approach for elements in an array, a repeating pulse signal, $v_A(t)$, can be represented by a single pulse convolved with a series of impulse functions separated at intervals T_{PRF}:

$$v_A(t) = g(t) * \text{III}(t/T_{PRF}) \tag{9.10}$$

$$v_A(t) = \prod(t/T_g)\sin(\omega_0 t) * \text{III}(t/T_{PRF}) \tag{9.11}$$

$$v_A(t) = \sum_{n=-\infty}^{\infty} \prod[(t - nT_{PRF})/T_g]\sin[\omega_0(t - nT_{PRF})] \tag{9.12}$$

where the *rect*, $\prod$, and the *shah*, III, functions were introduced in Section 6.2.1 for element samplers. From the time-frequency Fourier transform of this Eq. (9.12) and the application of the Fourier transform sampling property of the *shah* function, this spectrum results

$$V_A(f) = (-iT_{PRF}T_g/2)\{\text{sinc}\lfloor T_g(f - f_0)\rfloor - \text{sinc}\lfloor T_g(f + f_0)\rfloor\}\text{III}(f/f_{PRF}) \tag{9.13}$$

$$V_A(f) = (-iT_{PRF}T_g/2)\sum_{n=-\infty}^{\infty}\{\text{sinc}[T_g(nf_{PRF} - f_0)] - \text{sinc}[T_g(nf_{PRF} + f_0)]\}\delta(f/nf_{PRF}) \tag{9.14}$$

where the graphical representation of this equation is in the bottom of Fig. 9.7. A consequence of the repetitious life of the transmitted pulses is that their spectra appear as lines modulated by the *sinc*-shaped functions each centered on $\pm nf_0$.

From here the mathematics becomes much more involved and can be found in Szabo (2014b). Instead, the highlights of the analysis will be summarized.

To overcome range ambiguity, which is the well-known limitation of CW Doppler (which is sensitive to whatever vessels intersect its entire beam), pulsed wave Doppler (PWD) (Wells, 1969) was invented. Each transmitted pulse is affected by the flow it intersects and is Doppler-shifted, so that repeated pulses "sample" the flow as it moves in time.

An essential property of pulsed Doppler is that in addition to expected time dilation or contraction by the Doppler effect, changes in pulse arrival time are also involved (Jensen, 1996). For a transmitted waveform, $v_A(t)$, Jensen (1996) has shown that the received Doppler shifted output signal has the form (assuming no absorption or diffraction effects), of

$$v_B(t) = v_A\left[\psi\left(t - \frac{2d_0}{\psi c_0}\right)\right] = v_A\left[\psi t - \frac{2d_0}{c_0}\right] \tag{9.15}$$

where d_0 is the distance to the target, and the Doppler scaling factor ψ is

$$\psi = 1 - \frac{2c_s\cos\theta}{c_0} = 1 - \delta_D \tag{9.16}$$

that appears in Eq. (9.15) as a time scaling factor for dilation or contraction and also as a time delay modifier, and $\delta_D = 2c_s\cos\theta/c_0$, is an often-used constant that is very much like the constants used in the classic Doppler equation, Eq. (9.8). This Doppler scaling changes the *rect* part of each pulse from $\prod(t/T_g)$ in Eq. (9.11) to $\prod(\varphi t/T_g)$.

In CW Doppler, the Doppler-shifted received frequency is compared to the transmitted frequency; however, in range-gated Doppler, each received echo is compared to a similar echo resulting from the previous transmission. The relative delay between time-shifted Doppler echoes from consecutive pulses is simply (Jensen, 1996)

$$t_d = \frac{2\Delta z}{c_0} = \frac{2T_{PRF}c_s\cos\theta}{c_0} = \delta_D T_{PRF} \tag{9.17}$$

where Δz is the distance traveled away from the transducer. It is relatively straightforward to measure the time delays by cross-correlating successive pulses with each other to obtain an estimate of the flow velocity. Typically, a group of pulses is used to make a velocity estimate; such a group is often referred to as an "ensemble." Many more time samples of the flow are needed to calculate a fast Fourier transform (FFT) to obtain the Doppler spectrum which represents the range of apparent velocities attributed to the sample volume. The net result is numerically messy since both Doppler shifted and stationary frequencies are mixed together and need to be separated by filtering. This filtering is essential for measuring blood flow because the signal strength corresponding to the stationary scatterers (tissue, principally the vessel wall) is much

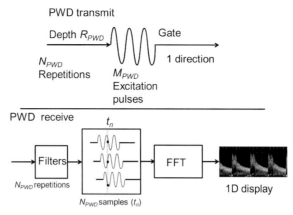

Figure 9.8 Conceptual diagram of timing and spatial sequences in pulsed wave Doppler. (Top) Parameters for pulse transmission. (Bottom) Parameters on reception leading to output display (Szabo, 2014e). *From Szabo, T. L. (2014e). Section 11.9. In Diagnostic Ultrasound Imaging: Inside Out (2nd ed.). Elsevier.*

stronger than the signal coming from the blood, often on the order of 40 dB stronger (Szabo, 2014d)! The filter must therefore do an excellent job rejecting the large component of the backscatter at low shift frequencies and must be adapted to the particular target; it is commonly known as the "wall filter."

The overall process is illustrated in Fig. 9.8. There are M_{PWD} transmitted pulses on a line and the line can be resent N_{PWD} times (N_{PWD} is the ensemble length, typically 32 to 64 acquisitions long). Typical values for M_{PWD} range from 4 to 12 cycles, with longer pulses providing better definition of the transmit frequency at the expense of spatial resolution, as discussed in Chapter 8. Filters remove the Doppler-shifted frequencies from the pulse echoes and the pulses are shown sampling the flow at different depths. Finally, an FFT is applied to the time samples to produce a Doppler spectrum. In addition, often an adjustable wall filter is applied which removes lower Doppler frequencies corresponding to adjacent tissue motion.

Aliasing occurs in the form of grating lobes for element periodicity exceeding half a wavelength; similarly, aliasing happens when the flow exceeds the sampling rate. Specifically, the flow velocity corresponding to a Doppler frequency f_D, must be below this Nyquist criterion:

$$f_D \leq f_{PRF}/2. \qquad (9.18)$$

If the flow exceeds this value, then inaccurate Doppler flow display or "aliasing" will occur.

PWD is relatively insensitive to the absorption and diffraction effects on the paths through intervening tissues to the target site. Pulse to pulse, these factors do not change much, so they are compared on a consistent basis; however, they affect overall sensitivity. Otherwise, absorption would cause a considerable downshift (on a Doppler scale) in the center frequency of the transmitted pulse (as discussed at the end of Chapter 3); consequently, it would generate a false Doppler signal (Jensen, 1996). Note that for the CW case, variations and loss caused by the diffraction of the beam and increasing absorption loss with depth can contribute to a diminishing sensitivity, which can be a problem for a real system with noise and limited dynamic range. Pulsed Doppler, when implemented on arrays, provides a number of advantages: a larger variable aperture, electronically controlled focusing and steering, and the ability to vary the sample volume by adjustment of the pulse length. More extensive pulsed Doppler information and theory can be found in Szabo (2014d).

The image in Fig. 9.9 shows a PWD display in combination with a color flow image which is a color-coded image of the overall flow velocities and will be explained in the next section. The green arrow points to the location of the Doppler range gate superimposed on the image. Note the two dots above and below it indicate the direction of the Doppler line. Below the image is the Doppler velocity spectrum in a display versus acquisition time. The range and interrogating pulse length and

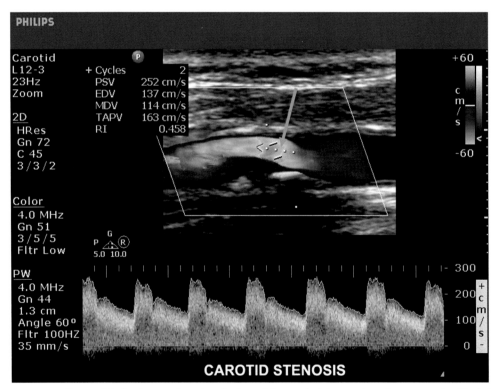

Figure 9.9 (Bottom) Pulsed wave Doppler display of flow velocities vs time for a stenosis. (Top) A color flow image showing placement of range gate, indicated by the green arrow. *Courtesy of Koninklijke Philips N.V.*

direction are controllable. From the image, an angle to the vessel can be determined and entered into some systems to correct for the Doppler *cosine* angle; in new systems the cosine correction is often applied automatically.

9.5 Color flow and power Doppler imaging

As shown in Fig. 9.9, color flow imaging is a way of visualizing flow velocities. In this figure, the speed of flow is color-coded as red toward the transducer (top of image) and blue away from it in cm/s units quantified by the color bar scale. Because this image contains three modes: color flow, B-mode, and PWD, it is known as a "triplex" image.

How is this image made? Once again, as in Doppler, repetitive sampling must be done over the same region to capture movement over time; the changes are small, and to detect them the timing and magnitude of the repeated pulses need to be very precise so as not to introduce noise. In addition, a B-mode image is also acquired to provide an anatomical background to position where the flow is happening. To accomplish this combined image, a region within the overall B-mode image is selected for flow imaging. An angle is also chosen for beam steering and calculations. Within the color flow region, several additional color flow lines are repeated in the same direction or position, then the next B-mode line is sent and the next group of color flow lines, and so on.

The process is illustrated in Fig. 9.10. For the color flow imaging (CFI) mode shown in Fig. 9.9, fewer samples are needed for velocity estimates than for PWD (spectral) waveforms. This region has a depth R_{CFI} (q) along a direction q and has a width defined by a number of lines, N_{lines}. A typical ensemble length (number of repeated samples) is $N_{CFI} = 8$ and a typical number of lines is $N_{lines} = 16$. Each ensemble of N_{CFI} points is moved in time for each line position identified by q in Fig. 9.10. Then the next position, $q + 1$, is processed to the maximum color flow depth, R_{CFI} (q), in the same way.

A frame rate can be calculated for an example that considers CFI as the only mode. If $R_{CFI} = 25$ mm, and the color flow image fills the whole area available with lines and/or blocks of lines, $N_{CFI}N_{lines} = 8 \times 16 = 128$. These numbers give a frame rate of 30 frames/s. With B-mode imaging added in, the frame rate drops to 27 frames/s; see Szabo et al. (1988). This example has an atypical shallow depth. If the depth is increased to 60 mm, the frame rate is 13 frames/s, and at 90 mm the frame rate drops to 8 frames/s.

Unfortunately, FFT processing is too slow with high uncertainty to be applied to the large imaging blocks used in CFI. An alternative method involves instantaneous frequency defined as

$$f_i = \frac{1}{2\pi}\frac{\partial \phi}{\partial t}, \tag{9.19}$$

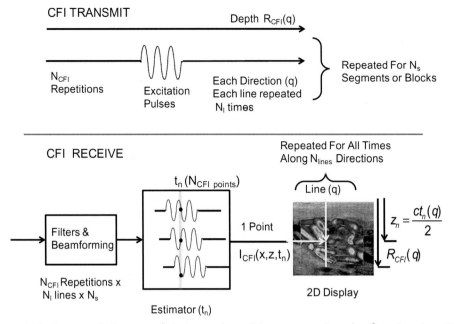

Figure 9.10 Conceptual diagram of timing and spatial sequences in color flow imaging. (Top) Parameters for pulse transmission. (Bottom) Parameters on reception leading to output display. *From Szabo, T. L. (2014e). Section 11.9. In Diagnostic Ultrasound Imaging: Inside Out (2nd ed.). Elsevier.*

where the signal as a function of time is expressed in terms of a magnitude, M, and phase, ϕ. Fewer points means that special faster algorithms other than FFTs with better signal to noise attributes need to be applied to make the best use of data. Although these algorithms such as discrete real-time autoregression estimators (Kasai et al., 1983) and others are beyond the scope of this discussion, details are summarized in Szabo (2014f). The output of these estimators is a color-coded point at z_n on line q in the color flow region of the image. (Actually, because time windows are used, several points in time are involved, but on the scale of the image, they appear as a pixel, $I(x,y,t_n)$.)

As in PWD, wall filters separate out the fluid flow velocities by removing the slower tissue movements. Alternatively, a high-frequency filter can remove fluid flow frequency components and keep and emphasize tissue motion in a presentation called "Color Tissue Doppler (Imaging)" or "Tissue Velocity Imaging." There is also a PWD counterpart, called "Tissue Doppler." The main applications of these modes are in echocardiography in tracking wall motion.

The essentials of a color flow image have already been presented in Fig. 9.9. They consist of a background B-mode image for context and flow placement and a highlighted color flow window in the form of a parallelogram, here tilted at a Doppler angle of 60 degrees. The B-mode image is formed at a center frequency of 7.2 MHz, whereas the color flow image (and pulsed Doppler) operates at a 4 MHz

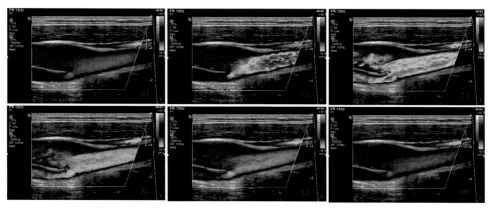

Figure 9.11 Color flow imaging sequence of a carotid artery bifurcation. *Courtesy of Koninklijke Philips N.V.*

frequency. At the upper right is the characteristic bidirectional colorbar indicating quantitative flow velocities and directions.

For another example of CFI, refer to Fig. 9.11. Here a sequence of frames shows the real power of this type of imaging: its ability to depict dynamic and complicated flow patterns. These images show velocities around a carotid artery bifurcation. Note the aliasing indicated by the presence of adjacent red and blue colors.

Finally, CFI has been implemented in three dimensions. An example can be found in Fig. 9.12.

9.6 Power Doppler imaging

A variation of CFI is called power Doppler, or ultrasound angiography (Babock et al., 1996; Bude & Rubin, 1996; Chen et al., 1996; Rubin et al., 1994, 1995), and is a color representation of Doppler intensity. Curious features of power Doppler are an absence of information about velocity direction and dependence on angle. If there is a sufficient Doppler signal, usually the presence of fluctuations of the backscatter with angle will be enough to show flow even at 90 degrees where $\cos \theta = 0$. What is being displayed is the integral of power density or

$$\int_{-\infty}^{\infty} P(f)df = \int_{-\infty}^{\infty} \left| V(f) \right|^2 df \approx \sum_{n=1}^{N} I^2(n) + Q^2(n) \qquad (9.20)$$

where P is power, V is velocity, and I and Q are the real and quadrature velocity signal components.

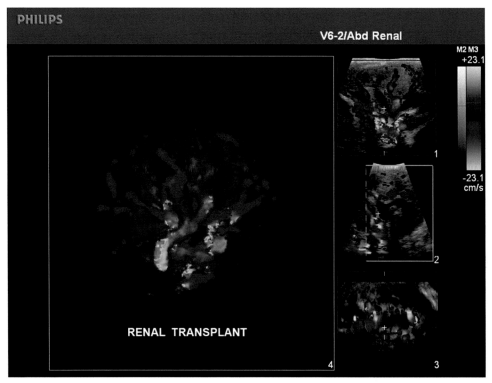

Figure 9.12 Three-dimensional color flow images of a renal transplant. A three-dimensional visualization is on the left with orthogonal views on the right. *Courtesy of Koninklijke Philips N.V.*

CFI and power Doppler imaging modes are compared in Fig. 9.13. Power Doppler imaging uses a color scale for intensity that goes from deep red to bright yellow, as indicated in the right pane of Fig. 9.13. In the power Doppler mode (Babock et al., 1996; Bude & Rubin, 1996; Chen et al., 1996; Rubin et al., 1994, 1995; Szabo, 2014f), the power of the Doppler signal without phase information is displayed as a range of monotone colors instead of directional colors. Noise arriving in the Doppler receiver at high gains is mapped to a small band of color in contrast to CFI, in which noise is spread across the spectrum as many colors. This containment of noise in the power Doppler mapping contributes to an effective increase in the dynamic range displayed. In CFI, fast moving tissue can overlap blood signals and appear as "flash" artifacts, but in power Doppler, this overlap appears as the same power amplitudes.

In the power Doppler mode, power is less affected by angular *cosine* effects; this advantage contributes to sensitivity improvement. Even at right angles, some signal is present because of the angular spread in beam formation. Another advantage is that aliasing at high velocities has little effect on the displayed power. These factors contribute to higher sensitivity for the depiction of small vessels. This improvement can be detected in the comparison shown in Fig. 9.13, for the smaller leftmost vessels.

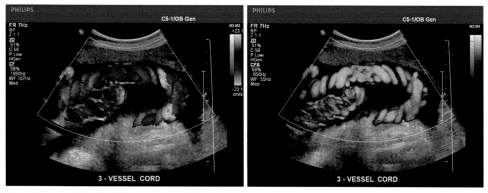

Figure 9.13 Left: Color flow imaging of umbilical cord. Right: Power Doppler image of same umbilical cord. *Courtesy of Koninklijke Philips N.V.*

9.7 Ultrafast Doppler imaging

PWD imaging includes more detailed quantitative information in the display than color flow Doppler because the spectrum of the Doppler signal is displayed over time, not just as a single velocity estimate (indeed, the term "spectral Doppler" is also used for this imaging mode). Obtaining this additional information requires collecting longer ensembles than those typically used in CFI. In line mode acquisitions, the Doppler gate is used to select the depth point along the transmit beam at which the PWD trace is to be collected. Multigate spectral Doppler allows multiple depth points along the line to be displayed as spectral traces, but these are confined to the same line because neighboring lines are not collected at the same time. Ultrafast Doppler imaging using plane waves allows for simultaneous collection of long ensemble data at all pixel locations within the beam, and permits spectral Doppler analysis at any set of points of interest in a complex flow, even those separated by a significant transverse distance (Bercoff et al., 2011).

"Ultrafast" imaging was originally developed to do shear wave elastography (Bercoff et al., 2002; Tanter & Fink, 2013), a method of launching shear waves inside a medium by "pushing" on it using focused ultrasound; see Section 10.5.2 for more details. Near the focus, the push deforms the medium using the radiation force, a force caused by a conversion of energy from the ultrasonic field to heat through absorption, resulting in momentum transfer from the field to the medium. The shear waves propagate in tissue-like media at a velocity of several meters per second, or several millimeters per millisecond. Since conventional line scan ultrasound can take tens of milliseconds to complete a frame, it is not able to capture the motion of shear waves. On the other hand, a plane wave beam can insonify the entire field of view in one transmission, and thus can image at kilohertz PRFs, that is, at a rate of several images per millisecond, capturing the shear waves as they propagate. Ultrafast imaging was then extended to high frame rate Doppler imaging (Bercoff et al., 2011), contrast bubble tracking, and other applications requiring very high frame rates.

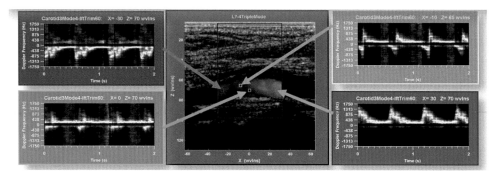

Figure 9.14 Multipoint spectral Doppler over a wide area using ultrafast plane wave insonification. The central panel presents a color flow Doppler image with the color overlaying a B-mode image; both images were acquired using plane waves. The four panels on the side present the spectral Doppler traces corresponding to the locations marked with tiny boxes. Traces can be obtained for any location within the Doppler region, delineated by the dark red box in the B-mode image.

As an example of using ultrafast imaging for wide area spectral Doppler, consider the multipoint PWD display in Fig. 9.14 (Flynn et al., 2012). The color flow Doppler image in the center panel presents a complex flow at the carotid bulb at a given instant in the cardiac cycle. This typical duplex display includes a B-mode image with a CFI overlay in a Doppler processing subregion delineated by the red box. In addition, several small markers have been placed in the flow region to obtain spectral traces for each one. These are updated in parallel, and it is easy to see that the flow characteristics are very different from each other. The traces were obtained by collecting data over two cardiac cycles and then replaying that data in a loop, with the time trace windows extending over four cycles, two of which are repeated. The ensemble length to create a single CFI frame is 12, whereas the PWD uses 64 pulse—echo acquisitions for a trace at a single time point. Additional pulse—echo acquisitions are collected to produce the B-mode image, for a combined PRF of 7000 Hz. With a PRF of 3500 Hz dedicated to the PWD alone, 55 spectral lines can be collected per second, though processing might not keep up in real time, and may need to be performed offline in a replay mode. Nevertheless, the advantage of ultrafast spectral Doppler is that the points at which spectral traces are requested are all simultaneously acquired, and such "gates" can be placed anywhere within the Doppler reconstruction region. Furthermore, the data can be recorded and analyzed at different times; the clinician can examine the flow in detail without requiring the sonographer to have placed the gate at the optimal location during the exam.

9.8 Vector Doppler imaging

9.8.1 Motivation

From the fundamental relation between the flow direction and the Doppler beam in Eq. (9.1), we know that the lack of a priori knowledge of the precise flow direction

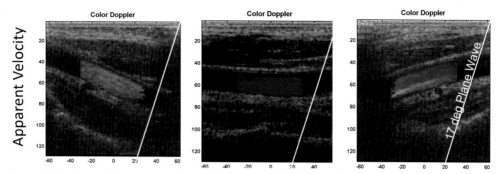

Figure 9.15 Three different color flow Doppler images for three different orientations of the transducer with respect to the vessel. Though each image measures the same flow (at the same point in the cardiac cycle) the color presentation of the flow is dramatically different, illustrating the dependence of the resulting image on the Doppler angle.

with respect to the transducer orientation results in a qualitative measure of flow velocity in a Doppler image: that is, the true angle θ is unknown. Indeed, even the direction of measured flow as represented by display color can appear differently depending on the orientation of the transducer and the beam direction, as illustrated in Fig. 9.15. The three images are all of the same vessel, and at the same point in the cardiac cycle, but it is clear that the Doppler image is very different because the angle between the transmit beam, here a plane wave, and the vessel is very different. While clinical systems sometimes provide automatic angle correction based on a good B-mode image of the vessel, these approaches are still heuristic and make assumptions that the flow will always follow the vessel axis. For this reason, vector Doppler imaging was developed to produce a measure of flow direction that is independent of the probe orientation. There are several approaches to vector Doppler imaging, and we will describe two here and others can be found summarized in Szabo (2014g) and Jensen (2016a, 2016b).

9.8.2 Combining Doppler information from multiple directions

One way to solve for the unknown angle θ between the flow and the Doppler beam is to take several measurements of the flow at different beam angles and fit the data to the $\cos \theta$ dependence. For line scanning, this may be cumbersome because of the large number of lines needed to cover a specific region of flow, and for this reason plane waves can be used because of the broad area they can insonify with a single transmission. Note that the PRF must be fast enough to measure the flow at several steered angles while the flow is essentially unchanged, and yet slow enough such that the flow is distinguishable from stationary tissue.

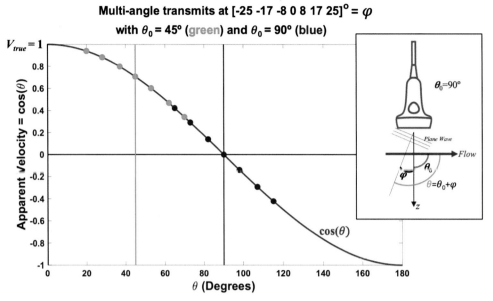

Figure 9.16 Fitting a cosine function (red line) to a set of Doppler measurements at different angles using a linear array. The blue dots represent Doppler velocity estimates at seven angles for a vessel that is perpendicular to the array axis and thus span the linear region of the cosine function; a large span of angles provides a good estimate of the true flow amplitude V_{true} (here normalized to 1). The green dots represent Doppler measurements for a vessel that is oriented at 45 degrees to the transducer axis and fall in the curved part of the cosine function. The inset illustrates the angle conventions used in the description.

How many angles are required? The approach is a curve fitting process with two unknowns: the true flow velocity amplitude V_{true}, and the angle between the beam and the flow, θ. Fig. 9.16 illustrates the challenge when the flow is close to perpendicular to the transducer axis, a common clinical situation. Here, the angle $\theta = \theta_0$ is the angle between the tranducer z-axis and the flow, and for $\theta_0 = 90°$ (blue dots) the cosine function is close to zero over the typical range of steering for a linear array with pitch on the order of 1 λ, and this results in uncertainty in estimating the flow velocity amplitude. Thus, using as wide a spread of steering angles is helpful in solving for the velocity direction *and* amplitude. For the range of plane wave angles (φ) used in Fig. 9.16 the angle $\theta = \theta_0 + \varphi$. The curve fitting solution is commonly obtained via a least squares approach. In addition to estimating true flow, vector Doppler provides information for those flows that are perpendicular to the transducer axis for which the axial component of flow is zero and is difficult to detect without moving the transducer or changing the Doppler beam angle.

While using more angles is beneficial to solving for V and θ, remember that an ensemble of acquisitions is needed for each angle, and using more angles leads to

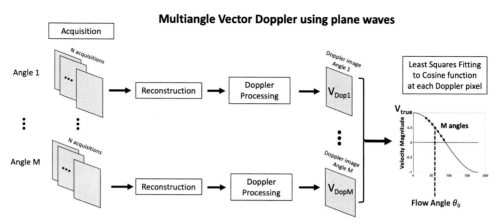

Figure 9.17 Multiangle vector Doppler data collection and processing scheme using plane waves and a linear array. Acquisition and processing proceeds for each angle as in color flow imaging, to create M images of $V_{Dop}(\theta)$. These M values are then fit to a cosine function to estimate the true flow velocity V_{true} and direction θ_0 at each pixel.

many more acquisitions. In principle, we can either interleave the ensembles, or collect them sequentially; the interleaved choice may not be practical if the PRF needs to be high to prevent aliasing, and it turns out that sequential approaches, in which the ensemble for one angle is collected together before moving on to another angle, have higher SNR.

Fig. 9.17 illustrates the acquisition and processing data flow for a multiangle vector Doppler sequence using plane waves. The ensembles of N acquisitions are collected one angle at a time, for M plane wave angles. Each ensemble is processed conventionally to provide the Doppler velocity estimate for that angle, V_a, that is, each acquisition is reconstructed, and the ensemble is wall filtered and processed to determine the flow velocity at each pixel location, forming the Doppler image for each transmission angle. The multiple angle data are then fit by least squares to a cosine curve, producing a velocity amplitude and direction at each pixel. Now two scalar quantities are associated with each pixel, and several different display approaches are used.

Fig. 9.18 provides an example of the results of a multiangle vector Doppler acquisition and processing scheme as described above. In this case, a 5 MHz linear array (ATL L7-4) was used to collect ensembles of $N = 12$ plane wave acquisitions at $M = 7$ steering angles (-25, -17, -8, 0, 8, 17, and 25 degrees), at a PRF of 4000 Hz, for a carotid artery and displayed here in a postsystolic phase for which the flow was quite stable. The display scheme uses color to display both the magnitude (top row, using the reddish half of the CFI color map) and the flow direction, by mapping flow angle to the color in a color wheel, displayed on the right (bottom row), with the 90 degrees flow angle mapped to purple. As can be seen, the flow magnitude is

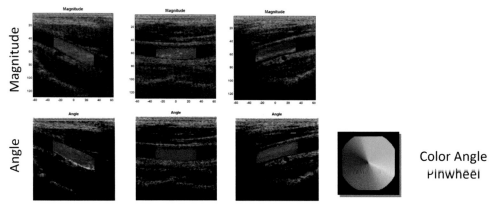

Figure 9.18 Vector Doppler, using multiple beam angles and solving for the true amplitude and flow angle. In the top row, the true amplitude of the flow is displayed, and is independent of relative orientation of the probe. In the bottom row, the flow angle is displayed as a color on the pinwheel. A nonuniform color across the vessel indicates that the flow direction is not constant, though here the carotid in the postsystolic phase exhibits nearly uniform flow in both magnitude and direction.

independent of transducer orientation, and the flow direction changes as expected for different angles between the transducer and the flow. The pinwheel angle origin is identified by a small tick mark, indicating that horizontal flow to the left is colored purple. Two other display methods will be discussed in the next section.

9.8.3 Vector Doppler using ultrafast plane waves from a single direction

Another approach to vector Doppler acquisition and processing is counter intuitive: using ultrafast imaging, it is possible to make vector Doppler measurements *using a single angle* (Goddi et al., 2017; Flynn et al., 2011). This is possible because at very high frame rates, the flow itself can be imaged in B-mode, that is, the speckle pattern produced by the blood can be frozen in a snapshot time. Furthermore, that speckle pattern can be tracked over time because the particles maintain their relative relationship to each other for a significant time to produce a highly correlated speckle pattern that is advected by the flow. One can use optical processing techniques from particle imaging velocimetry (Adrian and Westerweel, 2011; Raffel et al., 2007) to track the displacement of individual speckles from frame to frame on the spatial scale of a wavelength, that is, essentially at the pixel density. In doing so, the details of the flow, both vector direction and velocity, can be obtained in great spatial detail.

The results of this approach can be displayed using the color scheme of Fig. 9.18, but there are other display methods that can be more easily grasped by human perception. For example, one can render the flow as a cloud of point particles, each one displaced according to the local velocity vector. This visualization is much like the one used in the

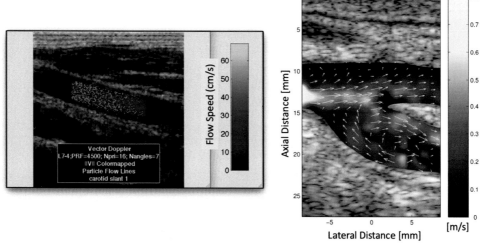

Figure 9.19 Two methods for displaying vector Doppler flow data: (Left) A single image from a movie in which particles in the flow field are assigned vector displacements between frames. The true velocity of blood is 50 times faster than the velocity used in moving the particles for visualization; the result is reminiscent of a flurry of snowflakes. (Right) An arrow plot in which the vector Doppler displacement estimates are rendered as a grid of tiny arrows on a uniform grid, with length proportional to the velocity amplitude and direction defined by the vector flow. Here, the arrows overlay a color rendering of the velocity magnitude. *Right image From Udesen, J., Nielsen, M. B., Nielsen, K. R., & Jensen, J. A. (2007). Examples of in vivo blood vector velocity estimation. Ultrasound in Medicine and Biology, 33, 541—548, used by permission.*

laboratory simulation of flow, as described in the next section. In practice, the flow visualized this way needs to be slowed down by a factor of 50 or so in order to be meaningful to our eye—brain processor. Such a method is best appreciated as a movie, but a still image from a movie generated in this way is presented in the left panel of Fig. 9.19.

Another vector Doppler single-direction real-time approach, the transverse oscillation method (Jensen, 2016a, 2016b; Udesen et al., 2007) uses arrow plots for display, in which tiny arrow markers are placed on a grid and indicate magnitude by their length and flow direction by their angle. These arrows are often placed over a color image of the flow magnitude; see Fig. 9.19, right panel.

9.9 The Vantage™ Doppler simulation using moving point scatterers

9.9.1 The flow model

Doppler experiments are usually carried out using phantoms that involve moving media, as simple as a string loop supported by pulleys and driven by a variable speed motor (JJ&A Instruments, 2023) to intricate flow phantoms in which a particle-loaded fluid is forced through tubes (CIRS Inc., 2023). While these are useful for study of physical motion and flow, and validation of Doppler processing algorithms, numerical simulations are also useful because the flow is known exactly and many flow configurations that may be impractical to achieve physically are easily developed. Because of these advantages, Verasonics has developed a flow model add-on to its general simulation tool that permits simulating motion in an acoustic medium by displacing scattering points between acquisitions according to user-defined algorithms. This enables a wide variety of flow models to be implemented. The basic approach is described in this section.

The Vantage software flow simulator is based on moving point scatterers and allows for flexibility in choosing simulation parameters, including the flow particles and the acquisition and processing sequence, at the expense of having to generate the synthetic radio frequency (RF) data from the moving scatterers for a number of frames. Here, the Doppler ensemble length is 12, and thus there are 12 acquisitions per Doppler image frame, not including the number of acquisitions used to create the underlying B-mode image. Using more frames takes longer to compute but produces a more interesting flow movie that illustrates the variability resulting from different random assemblies of scatterers.

The flow model uses a predefined density of scatterers, sufficient to create a reasonable (but not quite saturated) speckle field and yet small enough to be computable in a relatively short time: the models use on the order of a thousand points that are selected from a much larger 3D cylindrical model of the vessel, by assuming an elevation plane thickness of two wavelengths to represent the sample volume of the beam. The scatterers are of two types: tissue and blood, with the only distinctions being scattering (reflection) strength and movement. Scattering amplitudes for a given type are random within a dynamic range of a factor of about ten between the brightest and dimmest particles, that is, 20 dB. The mean tissue scattering strength is about 40 dB greater than the mean blood scattering strength.

Particle positions are random within the model. Only synthetic blood particles are displaced between transmits, according to a parabolic flow profile with minimum flow at the stationary wall, and the maximum velocity in the center corresponding to the user-specified flow velocity, as indicated in Fig. 9.20. The displacement amplitudes are also randomized, so that the blood particles move at slightly different velocities to mimic the motion of blood cells, even for the same radial position with respect to the center of the vessel. As a particle exits the edge of the model it is destroyed, and a new particle is created in its place at the entrance to the model.

The model is created out of cylindrical segments that can be concatenated to each other to form complex structures. The vessel segments can also be curved by applying a simple mapping of the particle positions according to a coordinate transformation. This approach leads to a variable density of particles for curved vessels, decreasing the density as the distance from the center of curvature increases, as can be seen in the semicircular model, in Fig. 9.20, right panel. This methodology does not generally create much of a problem, though the overall brightness of the vessel wall ("tissue") is

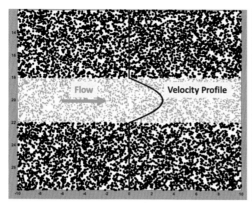

Figure 9.20 Vantage™ flow object moving point model representation in which each dot represents a particle in the model. The model includes stationary particles to mimic tissue, or vessel "wall" (black dots), and moving particles to mimic blood (pink dots). For illustration, the relative size of the dots indicates a difference in backscattering coefficient, here in the range of 40 dB. These images are frames from movies in which the motion is evident; here the flow direction is simply indicated by a green arrow. The flow velocity has a parabolic profile, indicated by the blue line; the specified flow velocity is assigned to the particles along the vessel center, and the actual velocity decreases toward the wall. A straight vessel (left) and a semicircular vessel (right) provide interesting flow geometries to explore with Doppler processing.

noticeably dimmer when the scatterer density is lower. Here, the particles are color and size coded to distinguish between blood and tissue types. The parabolic velocity profile is also illustrated by the blue line in Fig. 9.20, left panel.

These models use a relatively higher density and total number of particles ($\sim 10,000$) than the near real-time versions used in the lab and are only practical for offline simulation. The Vantage software initializes the particle positions for the selected model (straight flow, semicircular flow, branched flow) and then proceeds with simulated acquisitions, displacing the particles incrementally prior to each transmit, and then computing the RF backscatter from each particle to each element of the array to synthesize the simulated RF data.

9.9.2 Vantage Doppler imaging sequence using plane waves

The Vantage system can be programmed to use any sort of transmit beam desired; to minimize computation time, a single plane wave is used to insonify the entire set of particles at the expense of higher sidelobes. Each acquisition then comprises a full frame that is reconstructed into an image, with complex values at each pixel representing the demodulated RF in-phase and quadrature values, as $P = I + iQ$. The brightness of the pixel is given the magnitude $A = \sqrt{I^2 + Q^2}$, and the phase is given by the angle $\varphi = \mathrm{atan}\left(\frac{Q}{I}\right)$. As we have seen, B-mode processing takes the magnitude and compresses it before mapping to a grayscale for display. Doppler processing takes the ensemble of acquisitions, and for each pixel first removes the strong stationary (or slow-moving) signal component using a high pass filter called the "wall filter" because it greatly attenuates the echoes coming from the vessel wall and the sidelobes associated with those echoes. Then the processing seeks to estimate the phase change caused by the motion of the blood particles from frame to frame, averaged over the frames in the ensemble. This rate of change is converted to a velocity estimate and then mapped to a Doppler color scale forming a Doppler image.

How do we combine the Doppler image with the B-mode image in an overlay? The process requires some extra considerations to qualify the Doppler data and thus decide whether it is appropriate to show the Doppler or the B-mode at every pixel in the image. To this end, two additional parameters are defined: the Doppler power threshold (DPT), and the color write priority (CWP). The goal is to display the color data where there is true flow, that is, inside the vessel, and only when there is sufficient power in the Doppler estimate to assume that the estimate is due to real flow and not to sidelobes of the flow signal, as discussed in more detail below. The process of constructing a Doppler image on the Vantage is diagramed in Fig. 9.21. The approach is typical for "duplex" scanning, that is, interleaving acquisition and processing for two different imaging modes and then combining the results in one display. The duplex sequence is composed of two subsequences: a B-mode acquisition and processing procedure based on compounding multiple plane waves each at a different transmit angle and resulting in a single B-mode image frame, and a Doppler procedure in which N identical plane wave pulse−echo experiments are repeated, reconstructed, and processed to form a single Doppler image. The use of plane waves to produce both the B-mode and the Doppler images is less conventional.

The B-mode subsequence is a pixel-oriented processing pulsed plane wave coherent-compounding imaging approach, in which a number of planes waves are emitted at different angles (here, five plane waves are drawn) and the acquired RF data is reconstructed using a modified delay-and-sum beamformer for each frame. These reconstructed complex valued images are summed coherently, and the magnitude (or intensity in this diagram) is obtained from the sum to create a single B-mode image frame. As mentioned above, the frame data is logarithmically compressed and mapped to a grayscale color map for display.

The Doppler subsequence uses a single plane wave at a given angle in a pulse−echo experiment repeated N times, where N is the ensemble length. Each of the acquisitions is reconstructed as before, but now is processed by applying a wall filter to the ensemble (conceptually, independently for each pixel) and then cross-correlated to find the phase change per pulse repetition interval (PRI, where the PRF $= 1/$PRI). There are many different approaches to estimating the phase change between acquisitions over the ensemble, and further details can be found in (Evans and McDicken, 2000). The phase change or time shift is converted to a Doppler velocity (see Eq. 9.17) and this process results in the Doppler velocity estimate. The pixel velocities are mapped to a Doppler color scale as indicated in Fig. 9.21, where a typical Doppler color map is presented.

Plane Wave Doppler Imaging

B-Mode Subsequence

Doppler Subsequence

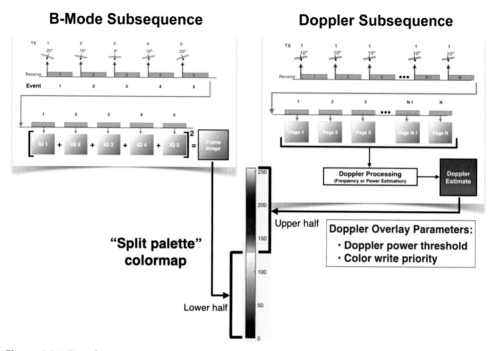

"Split palette" colormap

Upper half

Doppler Overlay Parameters:
- **Doppler power threshold**
- **Color write priority**

Lower half

Figure 9.21 Doppler imaging sequence using plane waves on the Vantage™ system. The Doppler color overlay onto the B-mode image is achieved by use of a split color map, in which either the grayscale or the color values are displayed depending on the values of the overlay parameters.

Note that the Vantage™ uses a split palette approach to render both the B-mode and the Doppler on one image. For each pixel, a decision must still be made as to which parameter should be displayed: the B-mode or the Doppler?

The Doppler power threshold is used as the first metric for qualifying the estimate: a low threshold would allow noise and sidelobes to appear, and a high threshold would eliminate all Doppler information. As a practical matter, the Doppler power is dependent on many parameters and is typically assigned a suitable threshold empirically. The students will find that the choice of threshold is very important and will experiment with it as they perform the simulations.

The color write priority essentially compares the intensity of the B-mode to the power of the Doppler at every pixel, under the assumptions that vessels will display as darker regions in the image and that the flow should be confined to vessels. One particular challenge is that successful use of the color write priority parameter depends on the quality of the B-mode image, specifically the contrast. Strong sidelobes and poor clutter reduction will reduce contrast and make it difficult to decide whether the flow estimate should overwrite the B-mode. This parameter is also adjusted empirically with the slider provided on the Vantage control interface. In spite of the relatively high sidelobes for plane wave B-mode images, suitable combinations of the two display parameters can usually be found for the simulations, except for regions outside the limits of the vessel model where the Doppler estimates are computed but the simulation has no meaning.

9.9.3 The Color Flow Doppler Simulator

The **Color Flow Doppler Simulator** creates a continuous flow movie with new particles flowing in from the left as old ones disappear on the right. The transducer is placed as usual, positioned above the image and pointing downward into the frame. The image displays flow in a conventional color scheme: red towards the transducer (warm colors, Fig. 9.22, right), and blue for flow away from the transducer (cool

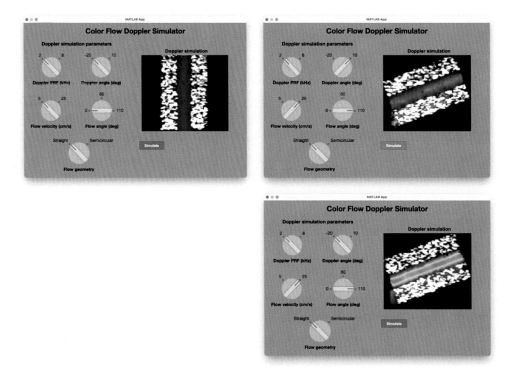

Figure 9.22 Doppler simulator examples using a point scatterer model. Stationary vessel wall particles produce the bright B-mode speckle, and moving particles produce the color flow Doppler image. (Top Left) The flow is moving down, away from the transducer, rendered in blue. (Top Right) The flow is moving upward, toward the transducer, and the parameters result in unaliased reddish results; color outside the model and inside the wall is an artifact of the underlying simulation. (Bottom Right) After changing the Doppler beam angle to better align with the flow direction (from +10 degrees to −20 degrees), the same flow used in the middle image is now aliased.

colors, Fig. 9.22, left). Vessel geometry can be either a straight or a semicircular tube. Both the flow angle (for the straight vessel only) and Doppler angle can be varied. The PRF and flow velocity in the center of the vessel can also be changed. The various flow options can be set in any combination. Note that for some combinations of parameters, aliasing may occur! The user can verify whether the flow should be aliased by evaluating the inequality of Eq. (9.18) and using the knowledge that the transmit frequency is $f_0 = 7.6$ MHz. Because the underlying simulation uses the parabolic flow profile of Fig. 9.20, a range of flow velocities is represented and an aliased case goes through the full range of flow colors moving from the wall toward the center of the tube, including some colors from the aliased color palette (Fig. 9.22, right). The simulator is very easy to use: click on the selector knobs to choose a set of parameter values, and then press "Simulate" to bring up that case.

9.10 Lab 9: numerically simulated Doppler imaging

9.10.1 Introduction

The Color Flow Doppler Simulator provides simple controls to select a limited set of values from key simulation parameters, including the flow direction (angle) and true velocity, the Doppler beam angle, and the Doppler PRF. The simulator is intended to rapidly provide a graphical view of CFI results to build intuition; the simulation is not intended to be faithful to physical flow dynamics, or even to be accurate outside the geometric limits of the underlying point scatterer model, even though a Doppler estimate may have been calculated in those regions. This is because, in the absence of stationary scatterers, the CPW algorithm has no suitable B-mode reference and assumes that flow is permitted. A specific subset of possible flow geometries and Doppler parameters has been preselected to produce a Doppler simulator that introduces CFI Doppler imaging. The students are instructed to explore all available parameter combinations to gain an appreciation for the type of display to expect.

Following familiarization with Doppler imaging using the Color Flow Doppler Simulator, the students can explore a more flexible numerical simulation on which the simple simulator is based. The **Vantage Doppler Simulation Tool** uses the point scatterer simulation capability built into the Vantage research system and provides the GUI shown in Fig. 9.23. This simulator allows exploration of all kinds of configurations, including some that are difficult to create with phantoms, and requires the students to use the various controls, including the CPT and CWP, to generate good images.

As mentioned in 9.9.1, the Vantage software includes a simulation capability that produces artificial RF data through the use of point reflectors which can be moved between acquisitions, like the approach used in the simulators of Chapter 8. The principal difference is that all of the postprocessing uses the same data flow as physically collected data would, allowing for tests of the complete programmed ultrasound sequence, including transmit and receive beamforming, transducer properties, and application postprocessing, especially including Doppler. A special set of functions has been developed to generate specific flow simulations including moving points with different scattering strengths to mimic stationary tissue and moving blood in a few interesting geometries.

9.10.2 Exercises using the Vantage Doppler simulation tool

The control interface for the Vantage Doppler simulation tool is presented in Fig. 9.23, left panel. The controls permit reproducing all of the cases selected for the Doppler Simulator, adding power Doppler imaging and an additional flow geometry. The students are first guided to reproduce some of those to learn how to use the DPT and the CWP parameters to optimize the image. Fig. 9.23, right presents the display for the default simulation settings at startup, after some adjustment to the default DPT and CWP parameter settings. Then, students may explore a wide range of flow conditions and Doppler settings, by selecting a combination of parameters. The program then runs the moving point acquisition sequence to create the RF data; and while the RF data is computed, the particles are plotted in a separate pop-out display window to show how they move in the model space from frame to frame, in an animation of images like those shown in Fig. 9.20. The sequence then enters a loop, repeatedly

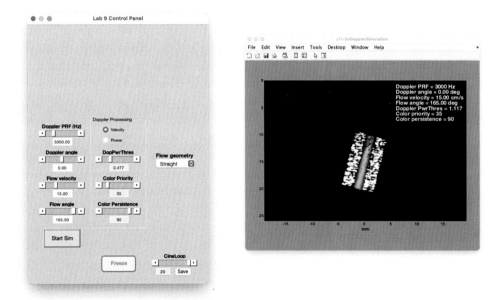

Figure 9.23 Vantage™ Doppler simulation tool graphical user interface for Lab 9 (left) and the display for the default starting parameters (right), after some adjustment of the Doppler power threshold and the color write priority to limit the Doppler image to the flow region.

processing that RF data and allowing the student to adjust the DPT and CWP image parameters as desired to optimize the display.

Finally, a third parameter is provided on the control panel to smooth the movie by adjusting the "color persistence." This parameter is commonly used to improve the appearance of Doppler by averaging several frames together using an adjustable weighting scheme. Thus, randomness is reduced, and a balance between responsiveness and artifacts can be established according to user preference.

Three flow geometries are provided for selection: a straight tube section, a semicircular tube, and a bifurcation. The straight vessel permits fundamental examination of flow and is a good model to use when beginning to explore use of the Doppler processing controls. The semicircular model provides a full range of Doppler angles in one view, and the bifurcation provides a few different wall thicknesses and vessel diameters to examine.

Semicircular flow model CFI results are presented in Fig. 9.24, showing an image that is just over the sampling limit of Eq. 9.18 (left) and an aliased image in which the Doppler frequency is significantly over the aliasing limit (right). The only difference between the simulations is the flow velocity: 15 cm/s on the left, and 25 cm/s on the right. The semicircular flow model is intriguing because it allows simultaneous visualization of constant flow in a counterclockwise direction over a 180 degrees span, coming toward (warm colors) and moving away from (cool colors) the transducer which is placed at the top of the image. Furthermore, though the flow is constant in magnitude around the semicircle, the color changes according to the cosine of the angle between the Doppler beam (vertical, 0 degree) and the flow direction; in particular, at 90 degree between the flow and the beam the apparent flow velocity goes through a null. Finally, the parabolic velocity profile is evident as a color variation at many locations along the vessel.

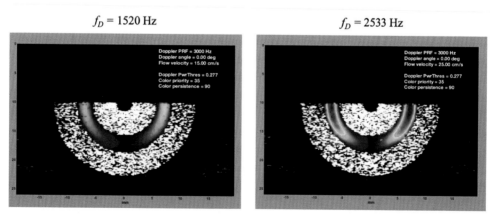

Figure 9.24 Semicircular flow model results, showing an image with $f_D \approx$ PRF/2 (left) and an aliased image with $f_D >$ PRF/2 (right), for a PRF of 3000 Hz. The only difference between the simulations is the flow velocity: 15 cm/s on the left, and 25 cm/s on the right. All Doppler simulations in this chapter use the transmit frequency $f_0 = 7.6$ MHz.

Students must also recognize that, in the interest of making the computation time as short as possible, these simulations are very rudimentary due to the relatively low particle density and simple acquisition modes. Speckle is not saturated, and coherently compounded multiangle plane wave imaging may result in dark "holes" in the B-mode image and strong sidelobes for both B-mode and Doppler. Nevertheless, seeing a point scatterer-based flow model turn into a simulated Doppler image in minutes is quite fascinating.

References

Adrian, R. J., & Westerweel, J. (2011). *Particle image velocimetry*. Cambridge University Press.
Babock, D. S., Patriquin, H., La Fortune, M., & Dauzal, M. (1996). Power Doppler sonography: Basic principles and clinical applications in children. *Pediatric Radiology*, *26*, 109−115.
Bercoff, J., Montaldo, G., Loupas, T., Savery, D., Mézière, F., Fink, M., & Tanter, M. (2011). Ultrafast compound Doppler imaging: Providing full blood flow characterization. *IEEE Transactions on Ultrasonics, Ferroelectrics, and Frequency Control*, *58*, 134−147.
Bercoff, J., Tanter, M., Chaffai, S., & Fink, M. (2002). Ultrafast imaging of beamformed shear waves induced by the acoustic radiation force. Application to transient elastography. In *2002 IEEE ultrasonics symposium, 2002. Proceedings: Vol. 2* (pp. 1899−1902). IEEE, Munich, Germany.
Bracewell, R. N. (2014). In *The impulse symbol. The Fourier Transform and its Applications* (3rd Edition, pp. 81−83). India: McGraw Hill Education.
Bude, R. O., & Rubin, J. M. (1996). Power Doppler sonography. *Radiology*, *200*, 21−23.
Chen, J.-F., Fowlkes, J. B., Carson, P. L., Rubin, J. M., & Adler, R. S. (1996). Autocorrelation of integrated power Doppler signals and its application. *Ultrasound in Medicine and Biology*, *22*, 1053−1057.
Chilowsky, C., & Langevin, P. (1923). *Production of submarine signals and the location of submarine objects* (United States Patent Office, 1,471,547). Application filed May 19, 1917, granted April 17, 1919. Serial No. 169,804.
CIRS Inc. (2023). *Doppler ultrasound flow phantoms*. https://www.cirsinc.com/products/ultrasound/zerdine-hydrogel/doppler-ultrasound-flow-phantom/. A subsidiary of Sun Nuclear, Mirion Medical Company.
Duck, F. A. (2022). Paul Langevin, U-boats, and ultrasonics. *Physics Today*, *75*(11), 42−48.
Evans, D. H., & McDicken, W. N. (2000). *Doppler ultrasound, physics, instrumentation and signal processing* (2nd ed.). John Wiley & Sons.
Flynn, J., Daigle, R., Pflugrath, L., & Kaczkowski, P.J. (2012). High framerate vector velocity blood flow imaging using a single planewave transmission angle. In *2012 IEEE International Ultrasonics Symposium proceedings* (pp. 323−325).
Flynn, J., Daigle, R., Pflugrath, L., Linkhart, K., & Kaczkowski, P.J. (2011). Estimation and display for vector Doppler imaging using planewave transmissions. In *Proceedings of IEEE ultrasonics symposium* (pp. 413−418).
Goddi, A., Bortolotto, C., Fiorina, I., Raciti, M. V., Fanizza, M., Turpini, E., Boffelli, G., & Calliada, F. (2017). High-frame rate vector flow imaging of the carotid bifurcation. *Insights Imaging*, *8*, 319−328.
Jensen, J. A. (1996). *Estimation of blood velocities using ultrasound*. Cambridge University Press, A recommended book for more details about Doppler and Doppler-related imaging and the measurement of blood flow.
Jensen, J. A., Nikolov, S., Yu, A. C. H., & Garcia, D. (2016a). Ultrasound vector flow imaging: I: Sequential systems. *IEEE Transactions on Ultrasonics, Ferroelectrics, and Frequency Control.*, *63*, 1704−1721.
Jensen, J., Nikolov, S., Yu, A. C. H., & Garcia, D. (2016b). Ultrasound vector flow imaging: II: Parallel systems. *IEEE Transactions on Ultrasonics, Ferroelectrics, and Frequency Control.*, *63*, 1722−1732.
JJ&A Instruments. (2023). *Mark 4 plus Doppler string phantom*. http://jja-instruments.com/Mark4Plus%20web.pdf.

Kasai, C., Namekawa, K., Koyano, A., & Omoto, R. (1983). Real-time two dimensional blood flow imaging using an autocorrelation technique. *IEEE Transactions on Sonics and Ultrasonics, SU-32,* 458−464.

Pierce, A. D. (1989). *P 454.* Acoustics. Acoustical Society of America.

Raffel, M., Willert, C., Wereley, S., & Kompenhans, J. (2007). *Particle image velocimetry: A practical guide.* Springer-Verlag.

Rubin, J. M., Bude, R. O., Carson, P. L., Bree, R. L., & Adler, R. S. (1994). Power Doppler ultrasound: A potential useful alternative to mean-frequency-based color Doppler ultrasound. *Radiology, 190,* 853−856.

Rubin, J. M., Adler, R. S., Fowlkes, J. B., Spratt, S., Pallister, J. E., Chen, J. F., & Carson, P. L. (1995). Fractional moving blood volume: Estimation with power Doppler US. *Radiology, 197,* 183−190.

Szabo, T. L. (2014a). *Overview. Diagnostic ultrasound imaging: Inside out.* (2nd ed., pp. 39−54). Elsevier.

Szabo, T. L. (2014b). *Section 11.2. Diagnostic ultrasound imaging: Inside out.* (2nd ed.). Elsevier.

Szabo, T. L. (2014c). *Section 11.4. Diagnostic ultrasound imaging: Inside out.* (2nd ed.). Elsevier.

Szabo, T. L. (2014d). *Section 11.5. Diagnostic ultrasound imaging: Inside out.* (2nd ed.). Elsevier.

Szabo, T. L. (2014e). *Section 11.9. Diagnostic ultrasound imaging: Inside out.* (2nd ed.). Elsevier.

Szabo, T. L. (2014f). *Section 11.7. Diagnostic ultrasound imaging: Inside out.* (2nd ed.). Elsevier.

Szabo, T. L. (2014g). *Section 11.10. Diagnostic ultrasound imaging: Inside out.* (2nd ed.). Elsevier.

Szabo, T. L., Melton, H. E., & Hempstead, P. S. (1988). Ultrasonic output measurements of multiple-mode diagnostic ultrasound systems. *IEEE Transactions on Ultrasonics, Ferroelectrics, and Frequency Control, 35*(2), 220−231.

Tanter, M., & Fink, M. (2013). Ultrafast imaging in biomedical ultrasound. *IEEE Transactions on Ultrasonics, Ferroelectrics, and Frequency Control, 61,* 102−119.

Udesen, J., Nielsen, M. B., Nielsen, K. R., & Jensen, J. A. (2007). Examples of in vivo blood vector velocity estimation. *Ultrasound in Medicine and Biology, 33,* 541−548.

Wells, P. N. T. (1969). A range gated Doppler system. *Medical and Biological Engineering, 7,* 641−652.

CHAPTER 10

Advanced ultrasound imaging systems and topics

10.1 Overview

10.1.1 Two views of ultrasound

Once, pulse-echo ultrasound was just an idea. Langevin made it a practical reality. Tasked with finding submarines underwater for the allies in World War I, Langevin and his team built the first practical long-range pulse-echo ultrasound systems which used piezoelectric crystals with the efficient crystal orientation we use today with directive beams, high-power transmitters, and high-gain electronic receivers (Duck, 2022, 2023).

From its beginnings, how did ultrasound grow and develop? Even though some have attempted to describe the chronological development of medical ultrasound (Szabo, 2014a, 2021; Goldberg et al., 2003; Woo, 2023), it is a long history that extends over 60 years, and it is a challenge to tell it completely.

Here instead, we look at the different multifaceted directions of ultrasound and where they are going. The growth of ultrasound can be charted as a tree. Even though this view is simplified, it has the value of identifying the main directions of effort in ultrasound. Secondly, the growth of ultrasound as a process is like the branches of the tree which sprout smaller and smaller new branches representing the increasing specialization we experience in our field. This representation also emphasizes the value of knowing the main branches of ultrasound so that the relationship between different parts of ultrasound science and engineering can be understood and expanded.

One view of ultrasound development shown in Fig. 10.1 is the American Institue of Physics (AIP) Physics and Astronomy Classification Scheme® (PACS) classification index. The organization of branches grew out of groups interested in certain topics rather than from an overall logical structure. Historically, Underwater Acoustics is one of the oldest branches with SONAR, underwater exploration, and the propagation of sound through water. Bioacoustics, as it relates to ultrasound, includes diverse areas such as ultrasound-induced bioeffects and neuromodulation. Nonlinear acoustics also has a long history and its own conferences and includes nonlinear propagation in water and air and harmonic imaging discussed later in this chapter. General linear acoustics covers the expanding knowledge of the fundamentals of acoustic wave and ray propagation including topics discussed in Chapter 2 and diffraction. The Physical Effects of Sound branch covers absorption, scattering, cavitation, and bubbles discussed later in

Essentials of Ultrasound Imaging
DOI: https://doi.org/10.1016/B978-0-323-95371-9.00008-4

Figure 10.1 Ultrasound knowledge tree based on Acoustical Society of America (ASA) classification scheme.

this chapter. Transduction includes ultrasound transducers described in Chapter 4. Structural acoustics covers nondestructive evaluation (NDE) and geophysical exploration with sound waves; both topics are introduced later in this chapter. Acoustical measurements and instrumentation would include imaging systems and methods of characterizing acoustic fields. Acoustic signal processing covers the block diagram approach in Chapter 2, Doppler processing in Chapter 9, and image processing in Chapter 8.

As of this writing, the Acoustical Society of America (ASA) publishes most papers in its journal, *JASA*, monthly; *JASA Express Letters* is a rapid dissemination online only journal; popular summaries of advances are found in *Acoustics Today*; and meetings are held twice a year. The text listing of the ASA PACS indexing scheme appears in Appendix B.

In the IEEE Ultrasonics, Ferroelectrics, and Frequency Control Society (UFFC-S), the Ultrasonics group has five subgroups. Group 1, for Medical Ultrasonics, is by far the largest and will be discussed later. Other topics of medical and other interests spill into the other major subgroups of the Ultrasonics group. Many papers are published annually in the UFFC-S transactions. Here are typical topics that fall within the interests of these groups and have been covered in the past annual International Ultrasonics Symposia.

Group 2 Sensors, NDE and Industrial Applications: acoustic imaging microscopy, general NDE methods, material and defect characterization, structural health monitoring, underwater acoustics, flow measurement, transducer energy harvesting.

Group 3 Physical Acoustics: General physical acoustics, nonlinear acoustics, acoustic tweezers, modeling, optoacoustics, thin films.

Group 4 Microacoustics, Surface Acoustic Wave (SAW), Thin Film Bulk Acoustic Resonator (FBAR), and Microelectromechanical Systems (MEMS): SAW modeling, MEMS and sensors, bulk acoustic waves (BAW) applications, materials.

Group 5 Transducer and Transducer Materials: materials and design and fabrication of large arrays, flexible and high-frequency arrays, Capacitive Micromachined Ultrasound Transducer (CMUT), Piezoelectric Micromachined Ultrasound Transducer (PMUT), integrated electronics for ultrasound imaging, transducers for neurostimulation and drug delivery, therapeutic transducers, multiwave and multimode transducers.

Already, some of the spillover topics covered in this book are included in these other groups and have been included or will be later in this chapter. Some examples include from Group 2: underwater acoustics and NDE (this chapter); from Group 3 physical acoustics (Rays and waves, Chapter 2, diffraction, Chapter 3); from Group 5 Transducers, Chapter 4.

The Group 1, Medical Ultrasonics, UFFC-S tree is shown in Fig. 10.2. Some topics may be familiar. Here, only one branch, contrast agents, is articulated into smaller branches otherwise the tree illustration would be overwhelmingly dense. Branches appear and grow quickly in the UFFC-S to keep up with the rapid changes in technology. A sampling of interesting topics on these main branches will be introduced later in this chapter and references will be given for further study. The list of UFFC group indices can be found in Appendix C.

10.1.2 In this chapter you will learn

This chapter is an introduction to advanced fascinating topics in ultrasound. While full details are not possible in the space allotted, references are supplied for further investigation. Maps are given in the first section to show the relationships among different topics and the process by which ultrasound grows branches of increasing specialization. Ultrasound imaging systems continue to evolve; the architectures of a commercial and a research system are explained. In general, tissues and materials are better described as nonlinear insofar as acoustic propagation through them is described. The characteristics of nonlinear propagation have been cleverly exploited through harmonic imaging and ultrasound contrast agents. Images of higher contrast have enabled better diagnosis through shear wave elastography. A hybrid method of imaging called "photoacoustics" enables the visualization of the optical properties of opaque materials with ultrasound. Higher-frequency ultrasound imaging reveals structures at very

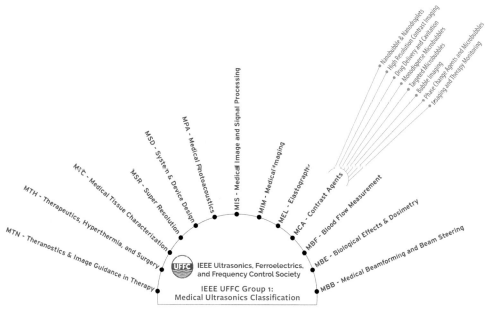

Figure 10.2 Ultrasound knowledge tree based on IEEE UFFC-S Classification scheme.

small scales. Three-dimensional (3D) imaging requiring either 2D arrays or mechanically scanned arrays in combination with visualization software provides unparalleled images of fetuses, internal organs as shapes, and tissue organization and function. Although it is not an imaging methodology, high- intensity focused ultrasound ([HIFU], and alternatively simply focused ultrasound [FUS]), this therapeutic approach can perform incisionless transcutaneous surgery, even at significant depth, by rapidly raising the temperature of tissues in tiny volumes similar in size to a grain of rice and deserves mention in this chapter. Super-resolution methods provide images of vasculature in exquisite detail. It has been used to advantage in two new areas of ultrasound application. One is functional ultrasound in which regions of increased blood flow in the brain are associated with external activities and thoughts and feelings from stimuli. Another is neuromodulation in which ultrasound directed at a selected region of the brain invokes some sort of activity, image, or movement. The advanced phased array technology used in diagnostic ultrasound can be applied to imaging problems in geophysics, material NDE, and underwater exploration.

10.2 Ultrasound imaging and research systems

10.2.1 Ultrasound imaging commercial systems

While the functional system block diagram has been explained in detail, its implementation has not. Fig. 10.3 captures the organization of a typical commercial imaging

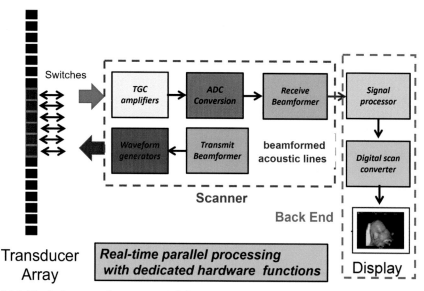

Figure 10.3 Block diagram of a commercial diagnostic ultrasound imaging system.

system. Delay-and-sum beamformers are by far the most common, but alternate beamformers are not ruled out. To handle the large throughput of processing needed for real-time imaging, special-purpose chips designed to operate in parallel are employed (Szabo, 2014b) to produce beamformed lines for the backend. Hidden from this system block diagram is the operation of one or more central processing units (CPUs), which guide and orchestrate the overall function of the system. These CPUs assume most of the back-end functions such as signal and image processing. In addition, specialized image modes such as Doppler imaging are processed there. As computer speeds increase, more processing ends up in software, and the research system discussed in the next section is the prime example.

10.2.2 Ultrasound imaging research systems

The Verasonics® Vantage™ Research Ultrasound System is a revolutionary, commercially available ultrasound system because it permits a single user to design, implement, and evaluate their own new ultrasound mode. This is possible because the system is an open programmable machine, permitting the user to configure all functions of the imaging system as diagramed in the block diagram, Fig. 3.1. The data acquisition hardware performs very limited functions, namely transmit and receive, digitization, and rapid data transfer to the host computer, and importantly, does no beamforming in hardware such as Application-Specific Integrated Circuits (ASICs). Because the Verasonics platform was initially inspired by the development of a complete software simulation of an ultrasound system, different architectural choices were made to permit new acquisition schemes and provide full access to the raw radio frequency (RF) channel data, features that are not possible, or very cumbersome, to accomplish with a conventional clinical system.

Fig. 10.4 illustrates the role of the hardware which transmits and receives the analog signals to and from the transducer array elements and stores the digitized data locally prior to transfer to the computer where all of the beamforming and signal processing occur. Even the switches are unconventional because the receiver is always connected to the transducer, and thus to the transmitter, with special circuitry in place to permit this arrangement.

An alternate view of the architecture is provided in Fig. 10.5. Again, the flexibility of the system lies in the fact that the ultrasound sequence of events is programmable and provides the developer the ability to specify: arbitrary transmit delays allowing for all kinds of transmit beams, receive properties with the flexibility of setting analog and digital properties and record length, time between acquisitions, and when to transfer data. The rest of the processing chain is implemented using parallelized and vectorized code that provides high-frame rate image reconstruction on a pixel grid using an efficient delay-and-sum beamforming approach. Application-specific postprocessing and image processing are also provided, but the developer can access the data at intervening points to implement their own algorithms at any stage in the data flow.

Verasonics' image reconstruction algorithm is a patented technology because it permits greatly reducing the number and complexity of computations usually required to form an image using conventional hardware-based beamformers. Called "Pixel Oriented Processing" (Daigle, 2017), it operates on the element-level RF data to create the image on a pixel grid, such as that illustrated in Fig. 10.6. The user specifies

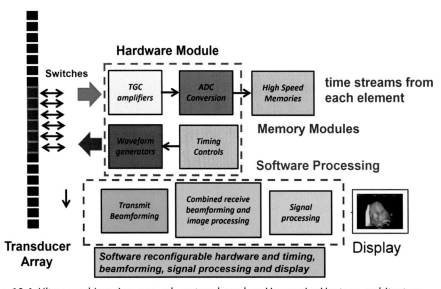

Figure 10.4 Ultrasound imaging research system based on Verasonics Vantage architecture.

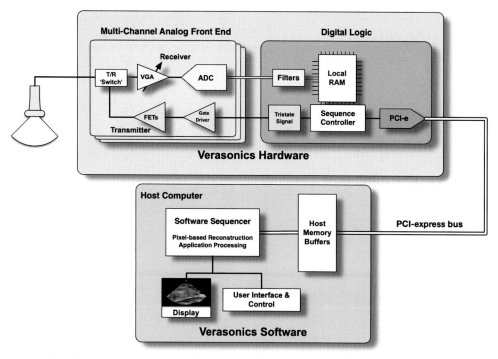

Figure 10.5 Block diagram of the Verasonics system architecture. The analog front end includes Transmit and Receive (T/R) and the Analog to Digital Converter (ADC). The digital hardware does no beamforming; it controls data acquisition and provides local memory storage (Random Access Memory – RAM) for data prior to transfer to the host computer where all of the beamforming and back-end processing and display are performed.

the location of the grid with respect to the transducer coordinate system and also specifies the extent and density of pixel locations where the reconstruction algorithm computes the phase-preserving demodulated data as a complex number representing the in-phase and quadrature components and designated as IQ data. This format is convenient because B-mode images can be easily assembled by taking the magnitude of the complex number, and Doppler processing is readily performed because the phase of each pixel value is easily computed. Pixels can be placed in any geometrical arrangement, though the most common are rectangular, as depicted in Fig. 10.6, or sector-shaped (polar), for use with curvilinear and phased array transducers as illustrated in Fig. 8.3.

The reconstruction process by delay-and-sum can be understood using the illustrations in Figs. 10.7 and 10.8. First let us review the creation of backscattered ultrasound data by examining the processes in a simple experiment diagrammed in Fig. 10.7. In the left panel, the linear array transducer is drawn as a red line at the top of the green rectangle that represents the field of view. Here, the medium is uniform except for some strong point scatterers drawn as black dots. The array emits a steered plane wave that first impinges on

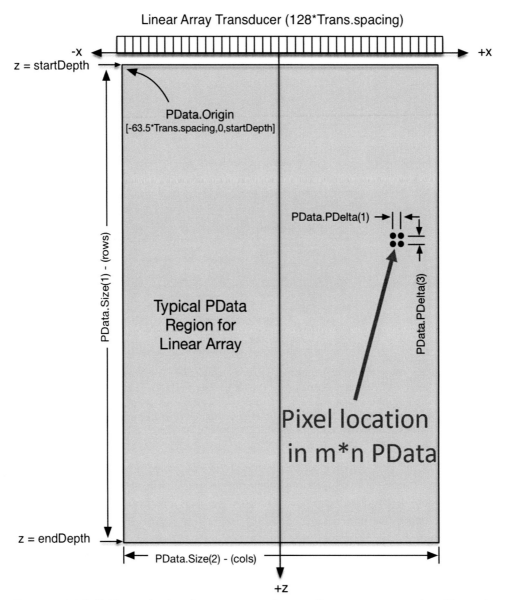

Figure 10.6 Definition of the image space for a linear array on the Verasonics Vantage Research System. Pixel location and density specifications appear in an $m \times n$ array called PData (pixel data).

the top four point targets, sonicating the rightmost point first, then moving left, the second, third, and fourth, in turn. Because they are small, these points then emit waves spherically in all directions, and those waves propagate back to the array where they

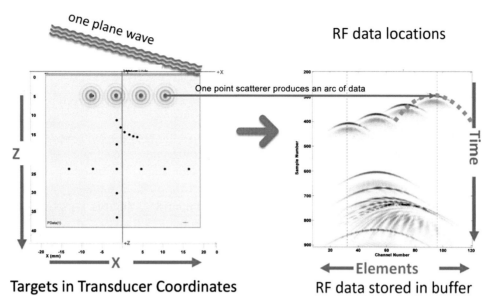

Figure 10.7 Left: Phantom target locations in image pixel matrix. Right: RF data corresponding to the echoes from the target locations in the top half of the image space are stored in the local system memory buffer in the Verasonics Vantage™. The data are then immediately transferred to computer memory for image reconstruction.

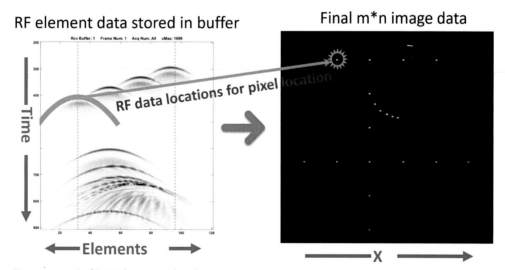

Figure 10.8 (Left) RF data stored in host computer memory are used to reconstruct the image at specific pixel locations for each pixel in the image (only the top half of the RF data for the complete image is shown). (Right) Data along an arc are time-aligned and summed to evaluate the pixel brightness. When the arc corresponds closely to echoes from a target, the pixel intensity is correspondingly bright.

impinge on the transducer elements at different times, with the shortest time corresponding to the distance to the array element directly above each point. The elements also have a directional sensitivity or directivity (Section 6.3) that leads to a weakening of the signal as the vertical angle increases. This leads to the RF data arcs in the right panel which presents a magnitude plot of the RF data for all elements versus time. Computing a simulation of this process can be called the "forward" problem, and the process of reconstructing an image from this data is an example of an "inverse" problem.

The reconstruction operation reconstitutes an image of the medium from the backscattered data and does so by applying a delay and sum procedure to compute each pixel's intensity in the image grid. For the image location corresponding to the leftmost point scatterer, as illustrated in Fig. 10.8, the data along the arc drawn in blue is summed to produce the pixel value of the reconstructed RF. To do the computation, the data from each element is aligned in time and then summed, including contributions from only those elements for which the element directivity value exceeds a programmed threshold. Because the data is sampled in time, interpolation is required in addition to time shifting. This procedure is the essence of delay-and-sum reconstruction, but the implementations of the approach vary in how the interpolation is done, how much of the arc is included, and whether any weighting is applied to the element data prior to summation. When the pixel location corresponds to a target location, and the medium sound speed that sets the curvature of the arc is correct, the pixel value is large in amplitude. When the pixel is not located on the target, the arc for reconstruction may still intersect with arcs in the data, and this results in image sidelobe artifacts. Various methods seek to reduce sidelobes, and several approaches are referenced at the end of the chapter. When successful, the reconstruction produces an image of the medium, but for the Delay-and-Sum (DAS) approach, the point spread function (PSF) is a diffraction-limited result with sidelobes. Distortions may occur due to using a reconstruction sound speed that is different from the true sound speed, and from heterogenous sound speed variations in the medium from the nominal value c_0 that are prevalent in tissue and in most natural media. The Vantage architecture enables researchers to replace the DAS beamformer with alternative algorithms that can reduce distortions and artifacts to improve imaging. If successful, such improvements can then be translated to clinical systems.

10.2.3 Comparison of ultrasound imaging systems

Commercially available diagnostic ultrasound imaging systems are compared to ultrasound research systems in summary form in Fig. 10.9.

Commercial diagnostic ultrasound imaging systems are designed for maximum clinical workflow efficacy and safe operation. To achieve successful clinical outcomes, the user interface is optimized for viewing, evaluating, annotating, and storing many images. Specialized default configurations are available for different clinical situations as well as appropriately designed transducer arrays.

Comparison of System Types

Commercial / Clinical	**Verasonics**
• FDA approved (safety regulations) • Simple hardware interface • Designed for clinical workflow • Limited flexibility • No access to internal data • Branded transducers only • No details on processing • Can only use pre-programmed modes • Mechanical index (MI)/ thermal index (TI) display (required bioeffects indicators)	• Not FDA approved • No hardware interface and no cart • Designed for research & development • Complete control of scan parameters • Complete access to internal data • Custom transducers • User defined processing • Permits creation of new modes • No mechanical index (MI)/ thermal index (TI)display

Philips EPIC Elite
ultrasound system

Verasonics Vantage
Research Ultrasound System

Figure 10.9 Left: Features of a commercial diagnostic ultrasound imaging system. Right: Features of an ultrasound imaging research system, the Verasonics Vantage™ system. Illustration by courtesy of Koninklijke Philips N.V. and Verasonics.

Clinical systems are highly regulated and conform to international standards and national safety requirements. As pointed out in Section 6.5.2, whenever the propagation medium is absorbing, heating will occur. Usually the heating effects from ultrasound by design are kept small. In the United States, an Output Display Standard was adopted (Lewin & Nowicki, 1998), and now, a continually evolving form of it has become part of international standards. In this approach, the system user is provided a relative measure of ultrasound-induced bioeffects as feedback during clinical practice in the form of specific bioeffect indices. The thermal index (TI) appears as an onscreen display of relative heating potential. Similarly, the mechanical index (MI) is used as a relative onscreen display of an estimate of the potential for ultrasound cavitation; the MI is based on the assumption that a bubble nucleation site may be present, which may or may not be the case. Options for reducing index values are to turn down the transmit voltage (if available) or switch to a lower index mode in keeping with ALARA principle (As Low As Reasonably Achievable). Because of safety concerns, every imaging mode is fixed by design and no access to the mysterious inner workings of the system is possible. Some commercial systems provide outputs in the form of digitized data.

Research systems have open software and so their operation is not controlled by safety regulations. Because the processing is done in software, it offers unlimited flexibility and access to system parameters. New transducers can be designed and configured to run on research systems. New modes can be created. However, if these systems are to be used to image or interact with people, some planning and regulatory approval are needed. Typically, the path may involve submission of an application to an institutional review board (IRB) and sufficient documentation that the clinical studies will be carried out with safeguards and the institution's clinical protocol. Recommendations for preclinical studies required for marketing new ultrasound imaging systems can be found on the USA Food and Drug Administration website (FDA, 2023).

10.3 Acoustic nolinearity and harmonic imaging

So far, we have assumed that materials and tissues are linear; however, in reality, they are nonlinear. Linearity means that proportionality and superposition hold and in general, linearity is a reasonable estimate. Just as some materials are more absorbing than others, they also vary in how nonlinear they are. Water is an interesting contradiction; its absorption coefficient is small but its nonlinearity is significant. Nonlinearity becomes noticeable if either the pressure (or particle displacement) or nonlinearity coefficient is large and then pressure at some field point is no longer proportional to the pressure amplitude at the source.

The difference between linearity and nonlinearity can be expressed mathematically (Szabo, 2014c). In a fluid, for example, the linear relation between variations in pressure and changes in density is described by Beyer (1997)

$$p - p_0 = A \left(\frac{\rho - \rho_0}{\rho_0} \right) = \rho_0 c_0^2 \left(\frac{\rho - \rho_0}{\rho_0} \right) \tag{10.1}$$

where p_0 and ρ_0 are the pressure and density at equilibrium in a fluid (similar relations can be found for gases and solids). Here, $A = \rho_0 c_0^2$ is a linear constant taken for $\rho = \rho_0$ and at a specific entropy. The assumption is made that the process is adiabatic, meaning that there is no heat transfer during the rapid fluctuations of an acoustic wave. Eq. 10.1 is a linear approximation to a nonlinear relation; a better approximation is to include a nonlinear term (Beyer, 1997) for the pressure as a parabolic function of density,

$$p - p_0 = A \left(\frac{\rho - \rho_0}{\rho_0} \right) + B \left(\frac{\rho - \rho_0}{\rho_0} \right)^2 + \ldots \tag{10.2}$$

called the "nonlinear equation of state," for state S, with B defined as

$$B = \left[\rho_0^2 \left(\frac{\partial^2 p}{\partial \rho^2} \right)_{S, \ \rho = \rho_0} \right] \tag{10.3}$$

From Eq. (10.3), it is evident that linearity holds as long as either the coefficient B or the change in density amplitude or their product remains small. A simple measure of the relative amount of nonlinearity is the ratio of B/A. More common, however, is the coefficient of nonlinearity, β, which is defined as

$$\beta = 1 + B/2A \qquad (10.4)$$

which will be related later to the speed of sound in a nonlinear medium. Coefficients for tissue fall in the range of 3–7 (water to fat) (Duck, 1990; Szabo, 2014d). Tissues are only slightly more nonlinear than water. Contrast agents (discussed in the next section) can have nonlinearity coefficients of more than 1000 in high concentration (Wu & Tong, 1997).

One of the fascinating characteristics of nonlinear propagation is that its effect is cumulative and waveform distortion becomes more pronounced with distance. In the top of Fig. 10.10 is a pressure waveform measured by an extremely wideband

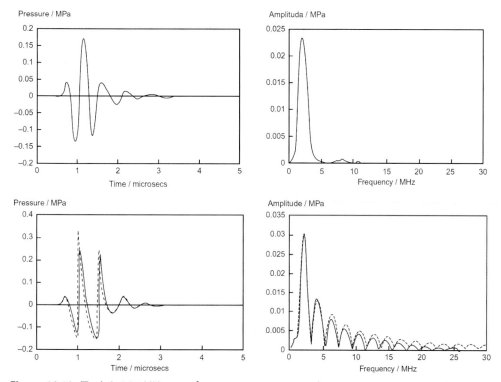

Figure 10.10 (Top) A 2.25-MHz waveform at source measured at $z = 15$ mm on-axis and used for simulation, with its spectrum on the right. (Bottom) Pressure waveform and spectrum simulated by KZK nonlinear propagation model and compared to data at 700 mm on-axis (Baker & Humphrey, 1992). *From Baker, A. C., & Humphrey, V. F. (1992). Distortion and high frequency generation due to nonlinear propagation of short ultrasonic pulses from a plane circular piston. Journal of the Acoustical Society of America, 92, 1699–1705.*

hydrophone transducer on the acoustic axis at only 15 mm from the transmitting transducer. This waveform is essentially linear and its spectrum is centered at 2.25 MHz as expected. At a distance of 700 mm, however, because of the nonlinearity of water, the waveform is severely distorted and its spectrum now contains a series of harmonics each centered at an integer multiple of the original center frequency. For example, the second harmonic is centered at 4.5 MHz.

One can take advantage of this harmonic generation feature of nonlinearity by filtering for just the second harmonic signal as illustrated in Fig. 10.11 (Duck, 2002; Humphrey, 2000; Simpson et al., 1999; Ward et al., 1996). The fundamental frequency emission, or first harmonic, must have sufficient power to generate harmonics due to medium nonlinearity, and this imposes requirements on the receive filter to reject transmitted energy, passing only the second harmonic. In the example of Fig. 10.11, the blue curve represents the L11−5v transducer's passband and is the two-way impulse response spectrum. The red curve is the transmit drive waveform spectrum centered at 5 MHz, which is modified by the nonlinear properties of the medium to add energy in harmonic bands, in particular, the second harmonic band around 10 MHz where the receive filter is designed (green line). Note that the higher harmonics which are also present in the medium are outside the transducer's passband and cannot be detected. This filtered signal can then be put to good use by increasing image resolution as demonstrated in Fig. 10.12 in an imaging phantom.

Imaging is commonly performed near the transducer's nominal center frequency, which corresponds to $f_c = 7.25$ MHz for the L11−5v transducer (center panel, Fig. 10.12). To find

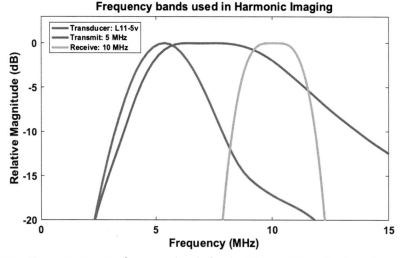

Figure 10.11 Harmonic imaging frequency bands for the L11−5v. When the transducer has sufficient bandwidth (blue) one can transmit at the lower end of the band (red) and receive in a band one octave above (green).

5 MHz
$f_{Tx} = f_{Rx}$

7.25 MHz
$f_{Tx} = f_{Rx}$

10.0 MHz
$f_{Tx} = 5\ MHz\quad f_{Rx} = 10\ MHz$

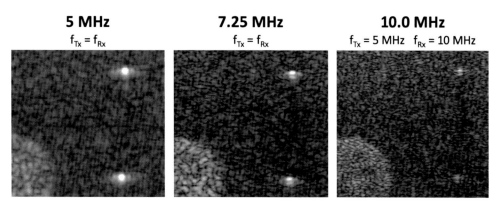

Figure 10.12 Using harmonic imaging to improve resolution in a conventional imaging phantom.

a pair of frequencies separated by an octave, the transducer's limited bandwidth requires lowering the transmit frequency to 5 MHz and receiving at 10 MHz. Of course, imaging at 5 MHz (left panel) would be worse than at 7.25 MHz but filtering and beamforming at 10 MHz produce noticeably improved resolution (right panel). Because the second harmonic signal is much weaker than the fundamental, significantly higher transmit voltages are usually required. Why not transmit and receive at 10 MHz? The attenuation for the round-trip path is lower for the harmonic imaging case because about half the path is weakly attenuated at 5 MHz and only the return trip is strongly attenuated at 10 MHz, in contrast to imaging using 10 MHz for both transmit and receive. Thus harmonic imaging permits improved resolution at deeper ranges than does transmit frequency imaging at the upper end of the transducer's passband.

The improvement in both axial and lateral resolution can be quantified by making profile measurements of point-like targets in the imaging phantom. Images of selected point targets are enlarged in Fig. 10.13 to show details and permit creating intensity profiles in axial and lateral directions. Imaging at the transmit frequency of 5 MHz results in a broad PSF (top row) when compared with the harmonic image obtained by filtering and beamforming at 10 MHz (bottom row). Nonlinearity can be explored in greater depth in Beyer (1997), Hamilton and Blackstock (1998), and Szabo (2014c).

10.4 Ultrasound contrast agents

10.4.1 Bubbles as nonlinear resonators

Contrast is the ability to distinguish something from a surrounding background. Because tissues tend to have similar impedances, reflectivity and contrast often have low values. Early investigators saw that air bubbles were bright reflectors in ultrasound images. These observations led to the deliberate injection of longer-lasting bubbles with more durable shells, filled with air or inert gases, that eventually were called "ultrasound contrast agents" (UCAs;

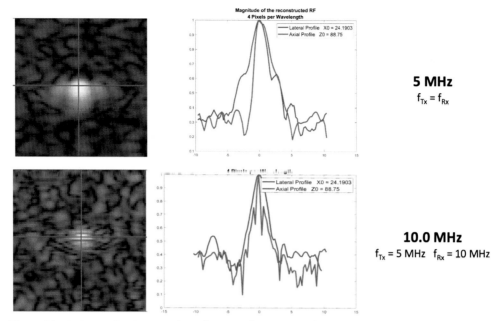

Figure 10.13 Comparing improved resolution with image profiles for a second harmonic image at 10 MHz to an ordinary 5 MHz transmit/receive image.

Szabo, 2014e). These bubbles had another useful property: resonance (Leighton, 1994, 1998). The bubbles expand and contract with the sinusoidal variations of the incident ultrasound as indicated in Fig. 10.14. Because it is physically harder to compress a bubble (compressional half cycle) than expand it (rarefactional half cycle), the bubble response is highly nonlinear as shown in Fig. 10.15.

Ultrasound contrast agents injected into the blood can highlight vessels and microvasculature and their nonlinearity has an even higher-contrast advantage. By bandpass filtering the second harmonic response, an even higher contrast can be achieved relative to the surroundings because the bubbles' nonlinearity coefficient is much larger than that of its surroundings.

10.4.2 Clinical applications of contrast agents

The enhanced contrast of UCAs imaged at the second harmonic is shown in Fig. 10.16, where images A and B produced at the fundamental lack the contrast evident in panels C and D. Other signal processing alternatives such as power modulation (panel E) and pulse inversion which subtract out the linear portions of the image are alternatives for achieving even more contrast. This type of brightening a blood region, here the left ventricle, is called "opacification."

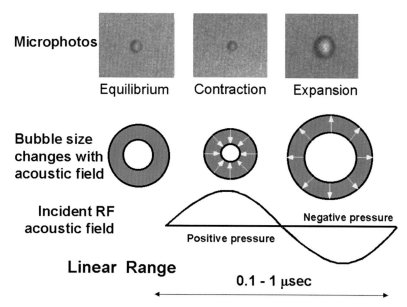

Figure 10.14 (Bottom) Symmetrical pacing of bubble expansion and compression with the compressional and rarefactional half cycles of an ultrasound wave. (Top) Measurement of this effect. *Adapted from P.G. Rafter, Philips Healthcare Systems.*

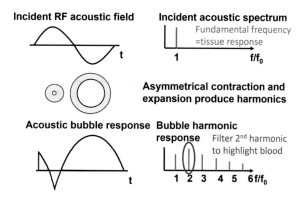

Figure 10.15 In higher-pressure incident sound fields, the microbubble response becomes nonlinear because compression is limited and shortened compared to expansion, leading to asymmetry and harmonics. *Courtesy of P. G. Rafter, Philips Healthcare Systems.*

The measurement of blood perfusing into muscle is a difficult one to make. The ability of the left ventricle to pump blood is related to how well it is nourished by blood perfusing into the heart muscle. With ultrasound contrast, myocardial echocardiography makes it possible to measure how quickly blood perfuses and to identify problems or diseased regions of the heart walls as illustrated in Fig. 10.17.

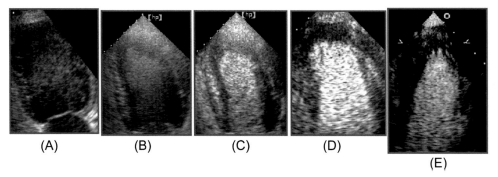

(A)   (B)   (C)   (D)

(E)

Figure 10.16 Evolution of ultrasound contrast agent imaging. (A) Imaging with first-generation agent at the fundamental frequency. (B) Imaging with second-generation agent at the fundamental. (C) Imaging with second-generation at the second harmonic. (D) Imaging with improved transducer field at the second harmonic. (E) Imaging with tissue-subtracting signal processing (power modulation). *Courtesy of P. G. Rafter, Philips Healthcare Systems.*

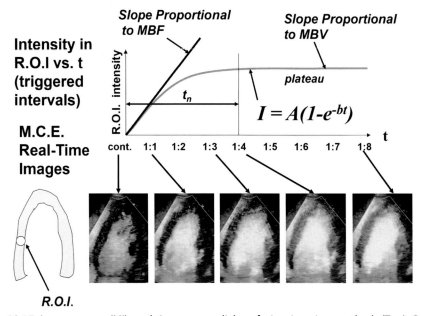

Figure 10.17 Low pressure (MI), real-time myocardial perfusion imaging method. (Top) Graph of region of interest intensity versus time perfusion filling curve, showing initial slope proportional to myocardial blood flow (MBF), a plateau region with a slope proportional to myocardial blood volume (MBV), and a time (tn) to reach the plateau. Time is in triggered-interval ratios such as 1:8, meaning an interval eight times the basic unit with reference to the initial administration of contrast depicted as "cont." (Bottom left) Insert highlights a region of interest (ROI) in the myocardium surrounding the left ventricle. (Bottom right) Time sequence series of left ventricle views depicting perfusion of the myocardium and beginning with contrast agent entering the left ventricle. *Courtesy of P. G. Rafter, Philips Healthcare Systems.*

Much more information about the applications of ultrasound contrast agents can be found in review articles by Cosgrove (2006), Powers et al. (2000), Qin et al. (2009), and Wilson and Burns (2010) and in the special issues on ultrasound contrast agents: Bouakaz and Dayton (2013) and Tang and Tortoli (2018).

10.5 Elastography imaging

10.5.1 Strain elastography imaging

As mentioned earlier, contrast between tissue types is often a problem because the longitudinal wave impedances of tissues, mainly composed of water, are about the same value, varying only by a few percent; consequently, reflectivity factors at tissue boundaries are low. Shear wave speeds of sound and elastic moduli for tissues, however, can be different by orders of magnitude (Sarvazyan et al., 2013).

How can we image shear elasticity to improve contrast (Szabo 2014f)? External forces can temporarily change tissue structure by pushing the surface downward to laterally displace tissue below and therefore create shear movement. Elastography originally was a quasistatic approach called "Strain Imaging" (Ophir et al., 1991, 2000). In this method, the precompression image of a region is captured with the transducer placed lightly on a surface. Next another image is taken with the transducer compressing the tissue downward in the same region. This vertical compression results in lateral displacements of tissue seen in the postcompression image. Tracking the speckle changes from the precompression image to the postcompression image reveals computed strains. The local strains show that hard places do not move much whereas softer, more compliant areas are more easily deformed. An example is illustrated in Fig. 10.18, where on the left is a conventional longitudinal pulse-echo precompression image of a hidden cyst with such low contrast that it is barely visible. The strain image in the inset on the right shows that the cyst is much softer than the surrounding material the subtle structure of which is also revealed.

10.5.2 Shear wave elastography imaging

Fortunately, there is another way of producing an almost pure, more sensitive shear elastography image and that is by generating shear waves inside the tissue and observing their motion. Shear waves can be generated from longitudinal waves directly, without externally pushing! These shear waves are generated as the byproduct of nonlinear propagation: absorption of the longitudinal waves leads to an acoustic radiation force (ARF). As illustrated at the top of Fig. 10.19, a transducer sends longitudinal waves into an absorbing medium. Some heat is produced along with a downwardly directed ARF. As the force is directed in the direction of propagation it displaces the tissue beneath it. These vertical displacements create slow shear wavefronts depicted in the figure. Shear wave speeds in soft tissues are only a few meters per second compared with about 1500 m/s for longitudinal waves. As

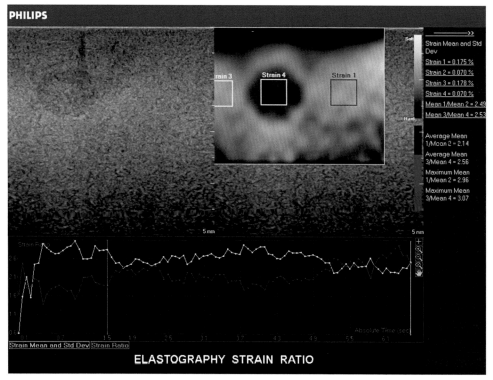

Figure 10.18 B-mode image of a cyst with a shear wave elastogram of the cyst in an inset. Strain ratios are taken midway through the cyst and laterally through the surrounding material below. The grayscale color bar indicates that the material is stiffer towards white and softer towards black. *Courtesy of Koninklijke Philips N.V.*

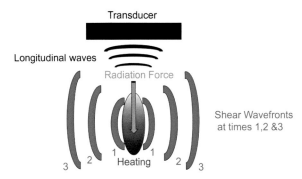

Figure 10.19 Interrelated ultrasound-induced bioeffects. A transducer sending downward directed longitudinal waves into an absorbing medium generates heat and axial acoustic radiation forces displacing tissue and causing vertically polarized laterally propagating shear waves.

the shear waves propagate through dissimilar tissues at vastly different relative speeds, their motion reveals improved contrast among the tissues. Stills from a shear wave propagation sequence taken on a Vantage system can be seen in Fig. 10.20.

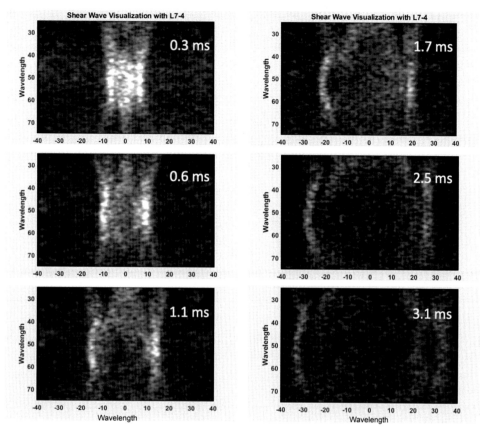

Figure 10.20 Experimental visualization of shear waves emanating from an acoustic radiation force impulse measured on a Vantage system. Further processing of these data can lead to a shear wave elastogram as in Fig. 10.21.

The frames were collected using broad beam "plane wave" sonications at a rate of 10 kHz, at the times indicated on each frame after the ARF impulsive push has ended. The push and the resulting image data were both produced using an ATL L7−4 transducer (128 elements, 5 MHz linear array) and driven by a Verasonics Vantage. The bright wavefronts represent vertical displacements propagating laterally. The images of shear waves in Fig. 10.20 were created by longitudinal waves sensing the vertical displacements caused by the rippling effect of the propagating shear waves; each shear wavefront image was generated by observing differences in phase between frames for a given location, that is, at each pixel. An inhomogeneity to the right of the push location visibly distorts the rightward propagating wavefront and generates a reflected wave from an invisible interface in the last frame. At the imaging frequency of 5 MHz, the longitudinal wavelength is about 0.3 mm, and the shear wave speed is thus approximately 10 λ/millisecond, or about 3 m/s, a factor of 500 times slower than the compressional sound speed! The longitudinal wave is fast enough to effectively freeze the motion of the shear wave in its position for each frame; only by comparing successive frames can the shear wave motion be obtained. This slow propagation speed is a direct consequence of the low stiffness of tissue-like media to shear (volume preserving) displacements, and the shear speed is used to determine the shear stiffness which has diagnostic value. A very high PRF is required to capture the motion of such shear waves when compared to typical frame rates. Such high frame rate imaging has been called "ultrafast" imaging, and image quality is typically sacrificed in favor of acquisition speed.

Recall the Elastic Wave Simulator and that shear waves were described in Section 2.3.3. Specifically, these are vertically z-polarized x-propagating shear waves represented by

$$\boldsymbol{u} = \hat{\boldsymbol{z}}\sin\left[(2\pi f_0)\left(t - x/c_S\right)\right] \tag{10.5}$$

where c_s is the shear wave speed. The motion depicted in Fig. 10.19 is not sinusoidal but like a highly damped Gaussian pulse (Sarvazyan et al. 1998), and investigators regard the speed of the wavefront to move at a shear wave speed. Because the position and timing of each image line, or alternatively, the timing and physical spatial locations corresponding to adjacent pixels are known, the local shear wave sound speed can be determined. This information can be used to create shear wave color maps or shear wave modulus maps from Eq. (2.18),

$$c_{44} = \rho c_s^2 \tag{10.6}$$

This is the basic principle, but because of absorption and dispersion (see Section 3.5.1) and other factors, implementations of Shear Wave Elastography (SWE) systems vary.

On the left of Fig. 10.21 is a B-mode image of a breast cyst and practically no information about the structure of the cyst itself. It is colored by a green overlay called a confidence map which signifies the reliability of elastography measurements in a region; a green color indicates high confidence in the data. On the right is the same longitudinal wave image but with a shear modulus map overlay. The colors correspond to quantitative values of the shear modulus related to Eq. (10.6) as indicated on the color bar. Alternatively, the shear wave speed can be displayed. The red regions indicate abnormally stiffer infiltrating tissue.

Several other implementations of elastography are of interest. An overview can be found in Szabo (2014f). Other reviews include Doherty et al. (2013), Fink and Tanter (2010), Greenleaf et al. (2003), Ophir et al. (1999), Palmeri and Nightingale (2011), Parker et al. (2011), and Sarvazyan et al. (2011).

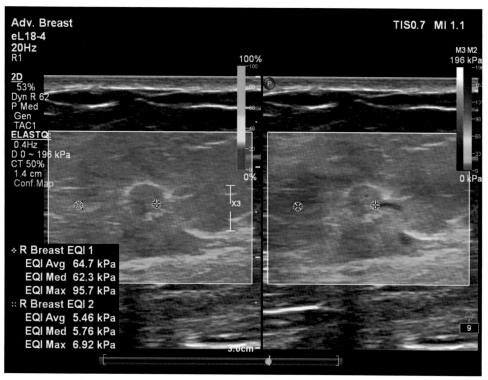

Figure 10.21 Left: B-mode image of a breast lesion with confidence map overlay. Right: a shear wave elastogram of the breast overlaid on a B-mode image. The quantitative shear modulus color bar indicates that the material is stiffer towards red and softer towards blue. *Courtesy of Koninklijke Philips N.V.*

10.6 Three-dimensional imaging

The key ingredient in making a 2D image is scanning across what we have designated as the x-axis. A-lines propagating, for example along the z-axis, are scanned across the x-direction to fill out an image. While the radiating side of a conventional array is physically a 2D surface, it only has a 1D array of elements, and, as we saw in Chapter 4, scanning can only happen in one dimension, along x. To produce a true 3D image, scanning in the other dimension, here y, is also needed (Szabo, 2014g). This double scanning is achieved electronically by a 2D array as discussed in Section 7.6. Examples of 2D or matrix arrays are given in Fig. 10.22. A number of scanning methods are available such as real-time fully populated arrays with microbeamforming which allow steering and focusing (Szabo, 2014h), row and column arrays (Section 7.6.2), annular arrays, and sparse arrays (Szabo, 2014i). A cheaper alternative is mechanically scanning a 1D array or annular array; however, some real-time capability is lost in this approach without advanced synchronization methods as described in Section 10.7.

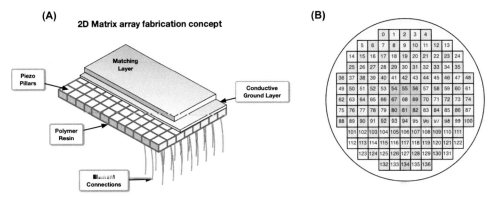

Figure 10.22 Matrix array construction. (A) Fabrication concept. Only three rows of wires are drawn for simplicity, and the backing layer is not shown. (B) Example of a circular aperture 137-element array, of which only 128 are connected in a checkerboard pattern to two banks of 64 transceiver channels. The central gray elements are not connected.

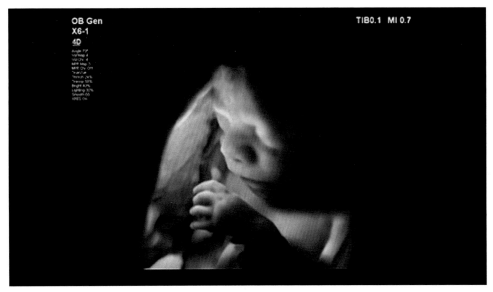

Figure 10.23 Three-dimensional surface rendered fetal image highlighted by virtually positioned light source. *Courtesy of Koninklijke Philips N.V.*

An example of a 3D image of a fetus is shown in Fig. 10.23. This is a surface-rendered image of a fetus with an artificial light source created in software to provide more realism. Perhaps even more amazing is that 3D ultrasound imaging captures the internal organs as well (not shown here but demonstrated in the following section).

10.7 High-frequency imaging

Higher frequencies result in finer resolution and detail in images. Achieving higher-frequency operation, usually starting above 20 MHz, requires specialized transducer fabrication including microminiaturization and low-loss materials. Because of high absorption, target regions are shallow and round-trip travel times are short, permitting very high-frame rate acquisitions which are often used with advanced signal processing to achieve adequate signal-to-noise (SNR). *In utero* or *in vivo* imaging also may place synchronization demands on data acquisition to avoid motion artifacts. Higher frequencies not only reveal fine detail as demonstrated in the left of Fig. 10.24, but because the scattering of small particles increases as frequency to the fourth power, blood (and its flow) is now directly visible to the left of the valve in the upper right two panels of the figure. Such details are better appreciated in the video from which these frames were collected (Verasonics.com, 2023).

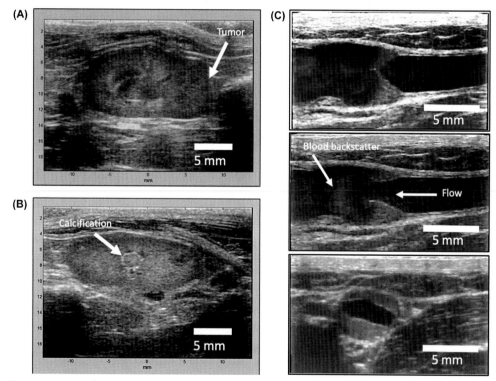

Figure 10.24 High-frequency imaging examples, using the Vantage system (high-frequency configuration). (A) Rat kidney, with 5 × 7 mm tumor, boundary indicated by the arrow (KOLO L22−8v CMUT transducer). (B) Rat kidney with calcification indicated by the arrow (KOLO L22−8v CMUT transducer). (C) Venous valve in closed position (top), open position with flow going from right to left (middle) and open, viewed transverse to the vessel (KOLO L38−22v CMUT transducer).

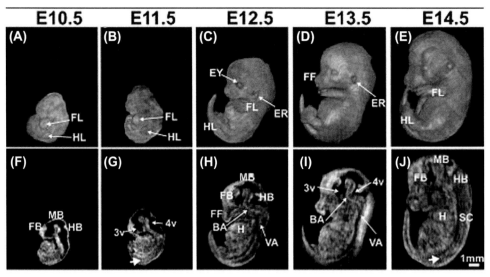

Figure 10.25 Whole embryo 3D *in utero* annular array high frequency imaging. Volume reconstructions (A–E) and the corresponding mid-sagittal sections (F–J) from array-focused data for the following embryonic stages: (A, F) E10.5; (B, G) E11.5; (C, H) E12.5; (D, I) E13.5; (E, J) E14.5. BA, basilar artery; EY, eye; ER, ear; FB, forebrain; FF, facial features; FL, forelimb; H, heart; HB, hindbrain; HL, hindlimb; MB, midbrain; SC, spinal cord; VA, vertebral artery; 3v, third ventricle; 4v, fourth ventricle. White arrows (G, J) indicate intersomitic blood vessels. Scale bar (J) = 1-mm. *From Aristizábal, O., Mamou, J., Ketterling, J. A., & Turnbull, D. H. (2013). High-throughput, high-frequency 3D ultrasound for in utero analysis of embryonic mouse brain development.* Ultrasound in Medicine & Biology, 39, *2321–2332, with permission from the World Federation of Ultrasound in Medicine and Biology.*

An example of 3D high-frequency imaging of embryonic brain development, Aristizábal et al. (2013), is shown in Fig. 10.25. On the top row are the surface-rendered external (volume) views of *in utero* embryos at days 10 through 14 of gestation. Below are the corresponding mid-sagittal planes revealing internal anatomy. These images were made with a 100% fractional bandwidth, 38 MHz center frequency, 6 mm *F#2* 5-ring annular array resulting in a spatial resolution of about 40 µm. Synthetic focusing and one hundred image planes spaced at 50 µm were acquired. Investigators employed advanced signal processing using a chirped-matched filter to improve signal processing and axial resolution. In each case the pregnant mouse was anesthetized, and placed on a heating pad and on a respiratory gating pillow to reduce motion artifacts with a tiny rectal thermometer inserted. An added complication was that typically a mouse mother would have three to four embryos.

10.8 Photoacoustics

Photoacoustics is a multimodal approach that combines optical contrast with ultrasound imaging. In Photoacoustics, a short energetic burst of light is emitted, illuminating the region

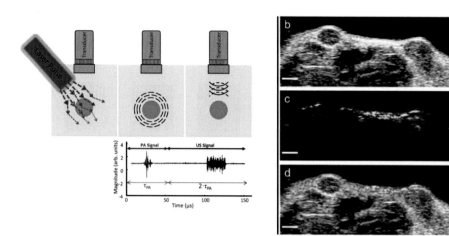

Figure 10.26 Photoacoustics uses powerful bursts of illumination to create sound by rapidly heating light absorbers. (left) Mechanism diagram, (b) ultrasound image, (c) photoacoustic image, and (d) overlay of photoacoustic image onto B-mode image (Bayer et al., 2012). *Reproduced from Bayer, C. L., Luke, G. P., & Emelianov, S. Y. (2012). Photoacoustic imaging for medical diagnostics.* Acoustics Today, 8, 15−23, with the permission of the Acoustical Society of America.

of interest. Regions with high absorbance at the light wavelength will heat during the burst and expand due to thermal expansion as depicted in Fig. 10.26A. This rapid mechanical expansion results in the emission of acoustic waves, with a frequency spectrum inversely related to the optical burst duration. These ultrasonic waves propagate outward and can be detected with an ultrasound transducer, as also illustrated in Fig. 10.26A. Ultrasonic absorption along the path will modify the emitted spectrum, as seen in Section 3.5.

The reconstruction of a photoacoustic image differs from the pulse-echo case because the light permeates the medium practically instantaneously with respect to sound. From rapid thermal expansion caused by light, all of the sources of sound emit at once. Because there is no transit time for the transmission, only a one-way receive ultrasound reconstruction is required.

The photoacoustic signal is rather weak, and many repeated acquisitions may be averaged to improve the SNR, though motion in the medium must be negligible over the averaging interval. Light intensity may be increased by using powerful laser sources, but average light intensity is limited by the FDA for patient safety, and the PRF for such sources may be limited.

Much like power Doppler imaging, the photoacoustic image is typically overlaid onto a B-mode ultrasound image as shown in the images on the right in Fig. 10.26. Commonly, a Power Doppler colormap is used, but any single-sided color map is appropriate. Frequently, red or near infrared light is used because it highlights regions where hemoglobin absorbs the light and produces high-resolution images of blood.

Optical contrast material can also be injected and it often absorbs over a specific narrow optical band, so that several different wavelengths of light can be used in turn to create photoacoustic images that depict the concentrations of different agents, allowing complex multicolor representations of the region of interest.

10.9 High-intensity focused ultrasound

HIFU also commonly referred to simply as FUS is not an imaging modality *per se* but is a very important medical application of ultrasound that deserves inclusion in this chapter. This method commonly refers to knifeless surgery for ablating unwanted tissue with precision. FUS has the advantage of reaching intact internal organs without incisions and without the normal attendant surgical side effects and longer healing time (Szabo, 2014j).

HIFU uses larger area transducers, often spherically shaped concave "self-focusing" devices, so that the intense field at the focus deposits enough heat in a small volume quickly so that it can quite literally "cook" tissue in the focal region while sparing surrounding tissue. This ablation can happen very rapidly, and it is localized to a region about the size of a −6 dB PSF ellipsoid, typically the size of a grain of rice. Steady-state heating due to absorption was covered in Section 6.5.3; HIFU ablation is typically due to a transient heating effect that produces coagulative necrosis and occurs at even higher nonlinear intensities in a short time (Szabo, 2014j). In contrast to ionizing beams, such ultrasound therapy can be reapplied to produce additional treatment as there is minimal damage to intervening tissue because of the high focal gains employed. Over time, researchers discovered that a wide range of acoustic regimes are useful and benefit from highly focused targeting, from very intense short pulses to long bursts at low intensity. In general, this area of ultrasound applications is called "therapeutic ultrasound" (Szabo, 2014k).

Fig. 10.27 provides a schematic view of noninvasive HIFU therapy, in which the transducer is coupled to the skin with gel or a couplant filled cone, and the sound traverses a number of tissue types before arriving at the targeted region. It is important that there should be an adequate acoustic "window," without intervening gas or bone which can simply block and reflect the passage of sound (gas) or block the beam and be heated undesirably (bone). Nevertheless, transcranial therapy is possible, using lower frequencies and very small $F\#$ transducers that are nearly hemispherical. The very high focal gain and relatively low frequency allow sufficient acoustic energy to pass through the skull, correct for skull aberration (Szabo, 2014l), and achieve successful and transformative ablative procedures in the brain, as is the case with a disease called Essential Tremor, and for which Focused Ultrasound Therapy, under magnetic resonance imaging (MRI) guidance, is FDA approved (Adams et al., 2021; Pajek & Kullervo Hynynen, 2012). Many other clinical applications are under investigation and are at various stages of development. The Focused Ultrasound Foundation provides excellent reports on the current state of the field, and advocates for the clinical

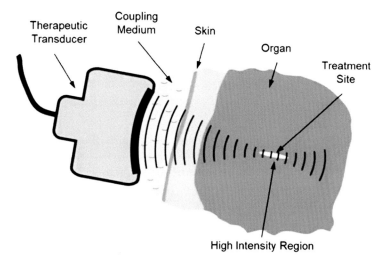

Figure 10.27 General diagram of a medical HIFU application. *Courtesy of Oleg Sapozhnikov.*

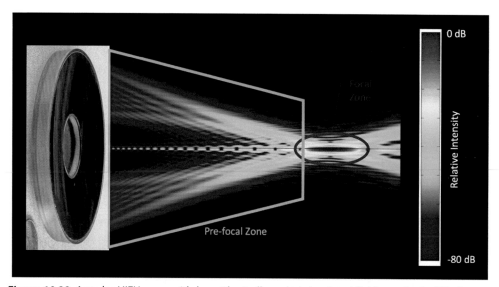

Figure 10.28 Annular HIFU array with logarithmically scaled simulated field map in the YZ-plane.

adoption of FUS therapies with high promise (FUSF, https://www.fusfoundation.org). In summary, when used with a suitable acoustic window and with suitable monitoring, noninvasive and highly localized therapy is achievable using ultrasound.

The use of a large aperture in two dimensions leads to enormous focal gain, which is proportional to transducer area as defined in Section 5.3.3. The axially symmetric beam pattern in Fig. 10.28 looks similar to the beams computed for 1D arrays, but the focusing gain is much larger because the diffraction-limited focal width is dependent on the *F#* and the frequency just as with the 1D array, but now the large aperture extends in both *X* and *Y* dimensions. For example, for an *F#*1 imaging transducer, the focal zone width is on the order of one wavelength, where the typical circular aperture of a HIFU transducer can be several times greater than that of an imaging array. The transducer photo in Fig. 10.28 has been distorted to give the appearance of proper perspective solely for the purpose of illustration.

Note that the transducer in Fig. 10.28 has a central opening. This is for placing an ultrasound imaging transducer, as also seen in Fig. 10.29 C and D, that can be used for guidance (positioning the HIFU transducer and planning focal zone placement) and monitoring of the therapy. It is essential to have ways of monitoring a noninvasive highly focused treatment, especially if it is intended to be ablative or if the target region is very close to other sensitive tissues. One challenge is that the imaging array typically needs to be placed near the HIFU transducer face so as not to obstruct the HIFU beam. This forces the imaging array to be much farther from the imaging region than normal, often resulting in poorer resolution and image quality than in typical imaging applications. While a detailed survey of monitoring methods is beyond

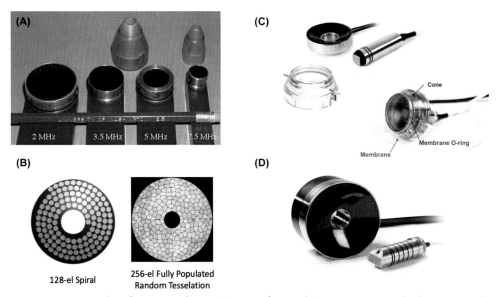

Figure 10.29 Examples of HIFU transducers. (A) Suite of surgical (intraoperative) single-element transducers and their coupling cones. (B) Spherical arrays with spiral packing, illustration courtesy of Sonic Concepts (left) and random mosaic tiling, image courtesy of Oleg Sapozhnikov (right). (C) Annular array of 65 mm diameter with central opening for a 5 MHz phased array imaging probe. A coupling cone is used to provide a standoff to use for shallow targets. The water-filled cone is sealed with a thin membrane. (D) 150 mm diameter spiral HIFU array with 128 elements and central opening for a 128-element 3.5 MHz phased array. The radius of curvature is equal to the aperture for an *F#* = 1.

the scope of this book, these methods are often novel ultrasound approaches that require an open, programmable research system to develop. Typically, the result of HIFU therapy is not easily visualized using conventional B-mode ultrasound imaging, and to monitor the effects of HIFU one needs new acquisition schemes and processing approaches to see and quantify progression of therapy.

For example, in ablative thermal therapy, the focal region is typically elevated beyond the tissue coagulation threshold, which for many tissues occurs within milliseconds at 65°C (Szabo, 2014j). Unfortunately, though the therapy has achieved its goal, B-mode imaging does not often produce a visible change. Tissue sound speed is a function of temperature, but it, and therefore the impedance, only changes by a very small amount, often much less than one percent. Nevertheless, it is possible to detect and image these minute changes by examining a time series of imaging frames interleaved with therapeutic exposures; the HIFU energy usually interferes with imaging so interleaving is generally helpful. Imaging small changes of sound speed or attenuation in the target region over time can produce useful monitoring capability. Other properties such as the shear modulus, tissue nonlinearity coefficients, and backscatter spectra can also be used to provide the "contrast" required to monitor therapy progression. The use of these properties and others is an active area of research at the time of this writing.

Fig. 10.29 provides several examples of HIFU transducers, from small, single-element intraoperative or shallow-target surgical devices, to large arrays, intended for deep targets in a transcutaneous approach. The devices are almost always self-focusing, because they use precisely machined piezoceramic shells of suitable radius of curvature and diameter, or diced composite material that is formed into a spherical shape. Manufacturers often specialize in one or the other fabrication technology, and each has benefits and disadvantages, so the choice of technology depends on the application.

Array designs permit steering the focus electronically over a region, but for 3D steering, the number of elements on the 2D surface grows large very quickly with the aperture and desired steering range. Fig 10.29B illustrates different array element designs. The goal with nonperiodic designs is to minimize grating lobe structure in the field. Grating lobes are inevitable given the relatively large element size, but random arrangements disrupt the coherent addition of energy and effectively lower the maximum sidelobe strength. Not shown are transcranial HIFU arrays employing thousands of elements which are needed not just to focus the beam but also correct locally for skull aberration (Adams et al., 2021; Pajek & Kullervo Hynynen, 2012; Szabo, 2014l).

10.10 Neuromodulation

One of the more exciting applications of focused ultrasound is its use in neuromodulation, that is, the stimulation of neural cells to either enhance or inhibit a particular function (Gavrilov, 2014). Fig. 10.30A illustrates the use of low-intensity focused ultrasound to inhibit

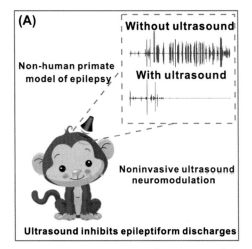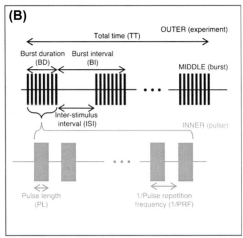

Figure 10.30 (A) Diagram of a neuromodulation experiment, in which the application of low-intensity ultrasound produces a physiological response, here inhibiting epileptic seizures monitored with EEG. (B) Typical ultrasound signal used to stimulate the nerve or brain region. *From Zou, J., Meng, L., Lin, Z., Yuan, T., Meng, W., Niu, L., Guo, Y., & Zheng, H. (2020). Ultrasound neuromodulation inhibits seizures in acute epileptic monkeys. iScience, 23, 101066, used with permission from Elsevier.*

epileptic circuits that generate unwanted signal activity (Zou et al., 2020). Brain treatments are commonly performed transcranially because the intensities used in neuromodulation are quite low and skull heating is less of a problem than it is for ablative procedures. Generally, the approach intends to preserve cells and prevent damage to them, and low-intensity ultrasound is used under a variety of exposure conditions to achieve instantaneous or long-lasting effects; see Fig. 10.30B. Ultrasound sequences often include several pulses of ultrasound at a given frequency, usually below 1 MHz for a particular pulse duration, and may be repeated at a PRF to make up one burst packet. Such bursts might also be repeated to constitute a treatment, which may also be repeated on a longer timescale, for example, successive days (Blackmore et al., 2019). See Fig. 10.31 for an illustration of a transcranial neuromodulation setup.

The field of neuromodulation is in its very early stages, and the mechanisms of action are yet to be elucidated. Once understood, treatment protocols will be better adapted than they are today to achieve the desired result. Several review articles may give the curious reader a place to begin learning more about this nascent and tantalizing field (Beisteiner et al., 2023; Blackmore et al., 2019; Bowary & Greenberg, 2018; Bystritsky et al., 2011; Gavrilov, 2014). Early work (Deffieux et al., 2013; Tyler et al., 2008) and selected references give a flavor of the approach to addressing specific applications (Butler et al., 2022; Lin et al., 2020).

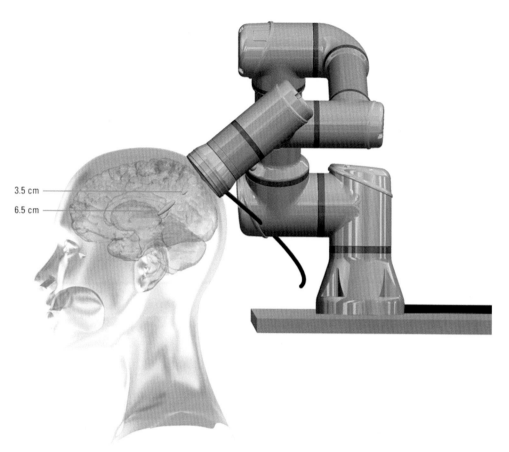

3.5 cm

6.5 cm

Figure 10.31 A robotic arm supports a focused ultrasound transducer and coupling system to treat a specific region of the brain. The annular array is able to change its focal depth electronically. *Courtesy of Sonic Concepts.*

10.11 Microvascular imaging and super-resolution

Super-resolution ultrasound (also known as ultrasound localization microscopy, ULM) is a method of achieving high-spatial resolution flow images of microvasculature (Errico et al., 2016; Song et al., 2023; Tang et al., 2020). The spectacular improvement in resolution (and sensitivity) jumps from hundreds of microns (for 5 MHz $\lambda = 300$ µm) to a few microns and provides the ability to see far smaller vessels, as illustrated in Figs. 10.32–10.34.

How is this possible? By injecting ultrasound contrast agents, which have a diameter of a few microns, into the vessels, as shown in Fig. 10.32, these microscopic point targets will not be resolved but rendered as blurry spots by the PSF of the ultrasound imaging system at each location as shown by the image on the right-hand side of the figure. Because ultrafast high-frame rate methods are used, each bubble position can be localized, and

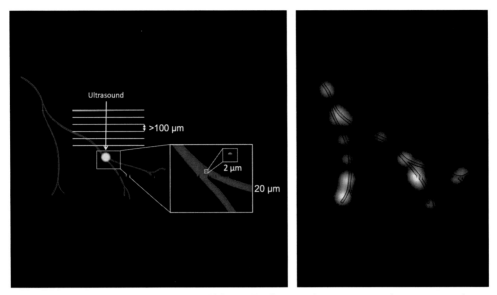

Figure 10.32 Tracking of dilute microbubbles with ultrasound to trace out the microvasculature. *Courtesy of Pengfei Song.*

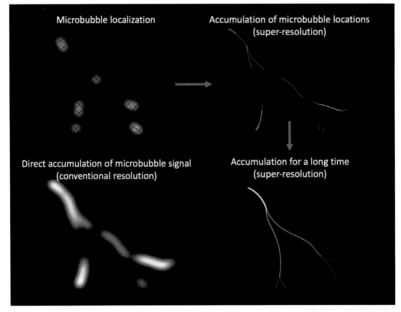

Figure 10.33 Localization of the center of individual bubble images in postprocessing enables super-resolved tracing of the microvasculature over time as the bubbles move with the flow. *Courtesy of Pengfei Song.*

Diffraction Limited   **Super Resolution**

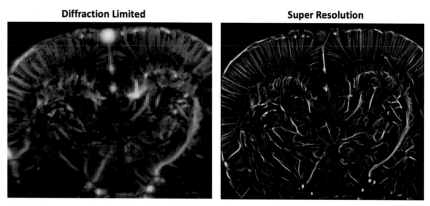

Figure 10.34 Super-resolution by microbubble tracking provides much greater detail in images of the microvascular flow and quantifies flow direction and volume. *Courtesy of Pengfei Song.*

individual bubbles are tracked to obtain the speed of each bubble. The cumulative aggregation of all the bubble locations over time fills in the vessels for imaging.

The expected results would look like the images on the left-hand side of Fig. 10.33. Blobs would appear at the bubble sites and over time they would trace out a blurry image of the vessel. However, since it is known that the bubbles are only a couple of microns in diameter and the blurry image of them is an artifact of the imaging system, the "localization" process replaces the PSF blob with a point and tracking determines the velocity of the bubble near that point. Then localization produces a much finer resolved vessel image as shown on the right-hand side. And over a long period of time (~ 150 s for 75,000 images), the bubbles "fill in" the rest of the spaces within the vessel to give a more complete image of the vessel shown in the lower right of Fig. 10.33. Finally, by tracking the movement of bubbles over time, the vessels are painted with colors representing the direction and speed of flow. The final results are illustrated in Fig. 10.34. On the left is diffraction–limited directional power Doppler image of a mouse brain. On the right side is the same brain but seen with the super-resolution method. A contrast-free super-resolution power Doppler technique has been developed that uses deep neural networks to achieve super-resolution with short enough data acquisition time to be implemented in real-time (You et al., 2023). For more background on this very active research area, see Couture et al., (2018).

10.12 Functional ultrasound

Functional imaging is a rapidly expanding part of neuroscience that can associate regions of the brain to cognitive activity and stimuli. Aside from increasing knowledge about how the brain works, it is also useful in revealing where and how regions of the brain change as a result of disease or to monitor recovery after surgery or other therapy.

The foremost approach is functional MRI (fMRI), an application of MRI to sense localized changes in brain activity, usually in the form of changes in cerebral metabolism, blood flow, volume, or oxygenation in response to task activation (Szabo, 2011). To emphasize these small changes, the preactivity data is subtracted from the postactivity data, and the resultant difference image is overlaid on a standard image. Special care must be taken when obtaining these images because the effect is small and can be corrupted by several sources of noise. Spatial resolution is on the order of mm and temporal resolution is in seconds and issues of noise and movement must be dealt with to achieve adequate SNR (Macé et al., 2013).

Recent advances in ultrasound imaging promise to dramatically improve spatial resolution and temporal resolution (Macé et al., 2013; Song et al., 2023; Szabo, 2014b). Since the introduction of ultrafast plane wave emission-based Power Doppler functional ultrasound imaging (PD-fUS) (Macé et al., 2013), an increasing number of studies are exploiting the capabilities of PD-fUS for functional brain imaging studies (Urban et al., 2015). However, the exact relationship between the PD-fUS signal and the underlying physiological parameters is quite complex and ultrasound Color Doppler (CD-fUS) is able to measure the axial blood flow velocity but suffers from unstable estimations of mean speed. The microbubble tracking-based ultrasound localization microscopy (ULM) method is able to map the whole mouse brain vasculature (coronal plane) and quantify the in-plane blood flow velocity with $\sim 10\,\mu m$ resolution (Errico et al., 2016). However, it suffers from a fundamental limitation of low temporal resolution as it requires extended data acquisition periods ($\sim 150\,s$ for 75,000 images) to accumulate sufficient microbubble events to form a single vascular image and corresponding velocity map, limiting its potential for functional brain imaging studies.

An alternative functional ultrasound method is illustrated here based on an ultrasound speckle decorrelation-based velocimetry method for blood flow velocity image of the rodent brain that overcomes the aforementioned temporal limitations and has diffraction-limited resolution (Tang et al., 2020). The setup of the functional ultrasound experiment is illustrated in the top left of Fig. 10.35 (Tang et al., 2020). An L22-14v and a Vantage system are used to acquire the data. The transducer is scanned along y to obtain whole brain data at one frame per second. A stimulation pattern that consists of a 30 s baseline followed by 10 trials of 15 s stimulation and with a 45 s interstimulus interval was used as shown in the upper right of the figure. The activation image, a1, shows the relative increase in blood flow in response to a left whisker stimulus. The flow versus time at the three locations is indicated in a2 and the relative increase in flow at these three positions is given by a3.

At the moment, functional ultrasound has been limited to animal studies but is advancing to work in humans. The main limitation is transmitting and focusing ultrasound through the skull. Impressive progress has been made in correcting for the aberrations caused by the nonuniformity and refraction through the skull for transcranial focused surgery using HIFU (Pajek & Kullervo Hynynen, 2012; Szabo, 2014l). While

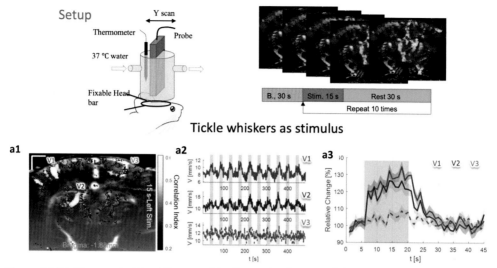

Figure 10.35 Changes in blood flow provide direct visualization of the regions of the brain associated with specific cognitive and motor activity using ultrasound speckle decorrelation-based velocimetry. Stimulus of a rat whisker leads to clear enhancement of flow observed over time in localized sites. Upper left: experimental setup; upper right: images in stimulus sequence; representative whisker stimulation results at Bregma approximately −1.88 mm. (A) Results of 15 s left side whisker stimulation; (A1) activation map; (A2) blood flow velocity time courses for the three vessels marked in (A1); (A3) 10 trials averaged relative response of the three vessels. Courtesy of J. Tang.

some progress is being made in transcranial imaging (Bertolo et al., 2021), studies are currently limited to animal models. For this reason, fMRI, which does not suffer from these limitations, is still the main functional imaging modality.

10.13 Material science: nondestructive evaluation/nondestructive testing

The science of materials evaluation and NDE and nondestructive testing (NDT) involve the measurements of the properties of solid objects, structures, joints, fasteners, rivets, and welds. Often data is taken by mechanically scanning these objects with single transducers in large water tanks to create 3D ultrasound images. Alternatively, measurements must be made in place where the parts to be inspected are in their operating environment.

Increasingly, phased array imaging systems speed up the process of characterization, as illustrated in Fig. 10.36. Matching layers increase the coupling to different materials such as metals, plastics, and ceramics. Shear waves and surface acoustic waves (SAWs) and noncontact transducers interrogate inaccessible locations and difficult geometries. Imaging requires more analysis and interpretation because of refraction through layers, reverberations, and propagation through composite structures. Most often scanning is used to locate and image defects, cracks, and material failures under load; data acquisition is combined with pattern or

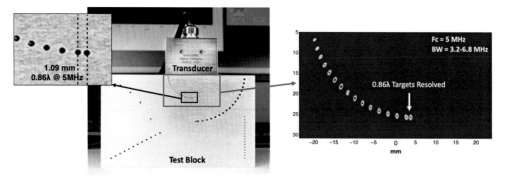

Figure 10.36 Ultrasonic imaging in NonDestructive Evaluation. (Left) Photograph of a steel calibration block with drilled holes that mimic voids, with the Vermon 5 MHz, 128-element linear array designed to couple to metal. (Right) Ultrasound image of the block using a reconstruction sound speed corresponding to the compressional velocity in the block.

flaw recognition software to automate the process. Of particular interest are the strength of joints and residual stresses that require nonlinear analysis. Objects undergoing different processes, loads, temperature or environmental changes, or assembly or production line operations pose interesting measurement challenges.

10.14 Underwater acoustics and SONAR

Underwater acoustics covers a vast area and many topics including military applications, fishing, ecology, seafloor prospecting, archeology, and biodiversity. A key difference in acoustic propagation in the ocean is that the speed of sound varies with depth z along thermoclines. A simple relationship for the angle θ of a sound ray propagating laterally is

$$\cos \theta(z)/c(z) = \text{constant} \qquad (10.7)$$

indicates that the angle $\theta(z)$ with the horizontal becomes smaller as c becomes larger or put another way, sound ray paths bend towards lower velocity regions (Kuperman & Roux, 2007). Sound travels great distances because of the low absorption of water and because of the channeling of sound. The surface of the water with the atmosphere acts as a rough mirror for sound and the fact that there is a sea bottom means the two boundaries create a duct or waveguide for sound. These rough boundaries scatter sound and can refract sound into the bottom, leading to attenuation. A far less lossy horizontal oceanic waveguide is often created by a sound speed minimum at an intermediate depth of typically hundreds of meters. This minimum is due to the combined effects of increasing sound speed with depth caused by hydrostatic pressure and increasing sound speed with temperature near the surface in the thermocline. This waveguide is called the sound fixing and ranging (SOFAR) channel and allows audible and subsonic frequency sound to propagate thousands of miles across the ocean! Much of underwater acoustics may sound familiar: there are transducer arrays, acoustic beams, pulse echoes, absorption, refraction, scattering, bubbles, ultrasound-induced bioeffects and imaging. Nevertheless, geometries are on a much larger scale with different problems.

Mapping the seafloor with a multibeam echosounder is illustrated in Fig. 10.37. This echosounder is a Konigsberg Simrad EM300 which has a 30 kHz array capable of transmitting 135 beams of a directivity beamwidth of one degree by one degree with a range from 10 to 5000 m. The scanning is done through the ship's forward movement at about 7—10 nautical miles/h. In order to cover a large area, the ship moves in a zigzag pattern with some overlap. As the ship is rocking with waves and wind, the GPS coordinates and timing are recorded so as to correct the direction of the beams, and their pulse echo returns by a receiving array. An example of an underwater survey of the Galapagos islands covered nearly 5000 km^2, the size of the state of Delaware, is given in Fig. 10.38 and could be called a hardwon C-mode image (Section 1.7.1). SONAR has been used to find lost cities, the Titanic ship, shipwrecks, and sunken planes as shown in Fig. 10.39. Subsurface information can also be obtained, the equivalent of an underwater B-mode scan as shown in Fig. 10.40.

Because water covers 70% of the earth, large regions of oceans remain unknown and are critical in determining weather and climate change. Sonic pollution such as constant noise from freighters carries great distances and disrupts underwater habitats. SONAR can also be lethal, especially to underwater mammals, so its safer use is promoted by international cooperation. Bubbles form a kind of acoustic noise in the form of strong scattering. The swim bladders of fish contain bubbles and bubble clouds occur naturally wherever waves break. The health of ocean life and the effects of warming waters are constantly studied and

Figure 10.37 Hull-mounted multibeam echosounder (University of Washington, Department of Oceanography Website, 2023). *Courtesy of the University of Washington, Department of Oceanography Website. (2023). http://ooicruises.ocean.washington.edu/enlighten/file/Hull-mounted + Multibeam + Echosounder.*

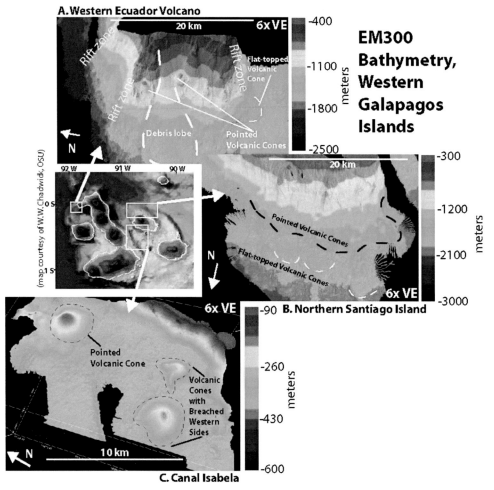

Figure 10.38 SONAR processed side-scan mosaic over one of the focused survey sites. Several prominent features are present in the mosaic that correspond with carbonate rocks and fish aggregations seen in the photographs taken by Sentry during the same survey. Light colors represent harder materials, while darker colors represent soft materials (i.e., sediment). *Image courtesy of DEEP SEARCH 2017, NOAA-OER/BOEM/USGS.*

monitored using acoustics, from coral reefs to the migration of whales. The search for subsurface minerals and resources is ongoing. Acoustics plays a key role in many of these underwater investigations.

10.15 Conclusion

As the ultrasound trees described in Section 10.1 indicate, the main branches develop into finer branches and increasing specialization. A common path for a scientist/

Figure 10.39 A high-resolution side-scan sonar image of a World War II B-25 airplane discovered in 2017 in Papua New Guinea. *Courtesy of Project Recover NOAA 2. (2023). https://oceanexplorer.noaa. gov/technology/sonar/side-scan.html. Accessed 21 May (NOAA 2, 2023).*

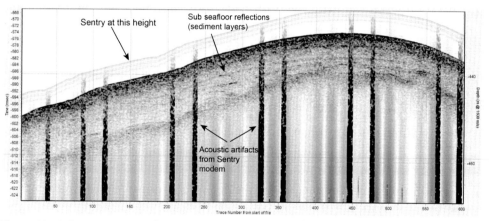

Figure 10.40 Screen capture of one of the preliminary processed subbottom profiles collected by the AUV Sentry during Dive 455. In this profile, subseafloor layers (stratigraphy) are visible as are several of the data artifacts that remain to be removed. Sentry transited at approximately 6 m above the seafloor when these data were collected. *Courtesy of DEEP SEARCH 2017, NOAA-OER/ BOEM/USGS, and National Oceanic and Atmospheric Administration (NOOA).*

engineer is to climb along one large branch to its end to be productive in the shortest amount of time. In this book, a different approach was taken. By beginning with basic principles, and gradually building in complexity, a foundation is built. Simulator examples graphically illustrate concepts dynamically and demonstrate which parameters are critical and identify physical constraints. Through the block diagram of an ultrasound imaging system, the interrelationships among different physical processes, system functions, and the signal flow are clarified. Once the physics of wave propagation, beam formation, and how transducers work and images are formed are understood, it is easier to learn about the new topics presented here. Though the applications introduced in this chapter are vastly different, there is a familiar ring to the approaches. For example in underwater acoustics, the use of arrays, transducers, beams, pulse echoes, absorption, refraction, scattering, bubbles, signal and image processing, and modes and ultrasound-induced bioeffects have all been discussed, for different applications. From the foundational principles introduced in this book and supplemented by simulators and laboratories, the new topics can be investigated further using the supplied references.

References

Adams, C., Jones, R. M., Yang, S. D., Kan, W. M., Leung, K., Zhou, Y., Lee, K. U., Huang, Y., & Hynynen, K. (2021). Implementation of a skull-conformal phased array for transcranial focused ultrasound therapy. *IEEE Transactions on Biomedical Engineering, 68*(11), 3457−3468.

Aristizábal, O., Mamou, J., Ketterling, J. A., & Turnbull, D. H. (2013). High-throughput, high-frequency 3D ultrasound for in utero analysis of embryonic mouse brain development. *Ultrasound in Medicine and Biology, 39*, 2321−2332.

Baker, A. C., & Humphrey, V. F. (1992). Distortion and high frequency generation due to nonlinear propagation of short ultrasonic pulses from a plane circular piston. *Journal of the Acoustical Society of America, 92*, 1699−1705.

Bayer, C. L., Luke, G. P., & Emelianov, S. Y. (2012). Photoacoustic imaging for medical diagnostics. *Acoustics Today, 8*, 15−23.

Beisteiner, R., Hallett, M., & Lozano, A. M. (2023). Ultrasound neuromodulation as a new brain therapy. *Advanced Science, 10*, 2205634.

Bertolo, A., Nouhoum, M., Cazzanelli, S., Ferrier, J., Mariani, J.-C., Kliewer, A., Belliard, B., Osmanski, B.-F., Deffieux, T., Pezet, S., Lenkei, Z., & Tanter, M. (2021). Whole-brain 3D activation and functional connectivity mapping in mice using transcranial functional ultrasound imaging. *JoVE-Journal of Visualized Experiments, 168*, e62267.

Beyer, R. T. (1997). *Nonlinear acoustics*. Acoustical Society of America, Reprint from Naval Ship Systems Command, 1974.

Blackmore, J., Shrivastava, S., Sallet, J., Butler, C. R., & Cleveland, R. O. (2019). Ultrasound neuromodulation: A review of results, mechanisms and safety. *Ultrasound in Medicine and Biology, 45*, 1509−1536.

Bouakaz, A., & Dayton, P. (2013). Special issue on ultrasound contrast agents and targeted drug delivery. *IEEE Transactions on Ultrasonics, Ferroelectrics, and Frequency Control, 60*(1).

Bowary, P., & Greenberg, B. D. (2018). Noninvasive focused ultrasound for neuromodulation: A review. *Psychiatric Clinics of North America, 41*, 505−514.

Butler, C. R., et al. (2022). Transcranial ultrasound stimulation to human middle temporal complex improves visual motion detection and modulates electrophysiological responses. *Brain Stimulation, 15*, 1236−1245.

Bystritsky, A., et al. (2011). A review of low-intensity focused ultrasound pulsation. *Brain Stimulation, 4*, 125−136.

Cosgrove, D. (2006). Ultrasound contrast agents: An overview. *European Journal of Radiology, 60*, 324−330.

Couture, O., Hingot, V., Heiles, B., Muleki-Seya, P., & Tanter, M. (2018). Ultrasound localization microscopy and super-resolution: A state of the Art. *IEEE Transactions on Ultrasonics Ferroelectrics, and Frequency Control, 65*, 1304−1320.

Daigle, R. E., (2017, May 16). *Ultrasound imaging system with pixel oriented processing* (US patent 9,649,094 B2).

Deffieux, T., Younan, Y., Tanter, M., Aubry, J.-F., Wattiez, N., & Pouget, P. (2013). Transcranial ultrasound neuromodulation of the contralateral visual field in awake monkey. In *2013 IEEE international ultrasonics symposium (IUS)*. IEEE, Prague, Czech Republic. https://doi.org/10.1109/ultsym.2013.0001.

Doherty, J. R., Trahey, G. E., Nightingale, K. R., & Palmeri, M. L. (2013). Acoustic radiation force elasticity imaging in diagnostic ultrasound. *IEEE Transactions on Ultrasonics, Ferroelectrics, and Frequency Control, 60*, 685−701.

Duck, F. A. (1990). *Physical properties of tissue*. Academic Press.

Duck, F. A. (2002). Nonlinear acoustics in diagnostic ultrasound. *Ultrasound in Medicine and Biology, 28*, 1−18.

Duck, F. (2022). Paul Langevin, U-boats, and ultrasonics. *Physics Today, 75*(11), 42.

Duck, F. A. (2023). Langevin's ultrasonic metrology. *IEEE Transactions on Ultrasonics, Ferroelectrics, and Frequency Control, 70*(2), 173−180.

Errico, C., Pierre, J., Pezet, S., Desailly, Y., Lenkei, Z., Couture, O., & Tanter, M. (2016). Transcranial functional ultrasound imaging of the brain using microbubble-enhanced ultrasensitive Doppler. *NeuroImage, 124*, 752−761.

FDA. (2023). Marketing clearance of diagnostic ultrasound systems and transducers. Available from https://www.fda.gov/regulatory-information/search-fda-guidance-documents/marketing-clearance-diagnostic-ultrasound-systems-and-transducers. (2023) Accessed 21.05.23.

Fink, M., & Tanter, M. (2010). Multiwave imaging and super-resolution. *Physics Today*, 28−33.

Focused Ultrasound Foundation https://www.fusfoundation.org.

Gavrilov, L. R. (2014). *Use of focused ultrasound for stimulation of various neural structures*. Nova Publishers.

Goldberg, B.B., Wells, P.N. T., Claudon, M., & Kondratas, R. (2003). *History of medical ultrasound*. A CD-ROM compiled by WFUMB History/Archives Committee, WFUMB, 10th congress, Montreal. A compilation of seminal papers, historical reviews, and retrospectives.

Greenleaf, J. F., Fatemi, M., & Insana, M. F. (2003). Selected methods for imaging elastic properties of biological tissues. *Annual Review of Biomedical Engineering, 5*, 57−78.

Leighton, T. G. (1994). *The acoustic bubble*. Academic Press.

Leighton, T. G. (1998). An introduction to acoustic cavitation. In F. A. Duck, A. C. Baker, & H. C. Starritt (Eds.), *Ultrasound in medicine, medical science series*. Institute of Physics Publishing.

Hamilton, M. F., & Blackstock, D. T. (Eds.), (1998). *Nonlinear acoustics*. Academic Press.

Humphrey, V. F. (2000). Nonlinear propagation in ultrasonic fields: Measurements, modelling and harmonic imaging. *Ultrasonics, 38*, 267−272.

Kuperman, W., & Roux, P. (2007). Underwater acoustics. In T. Rossing (Ed.), *Springer handbook of acoustics*. Springer. Available from https://doi.org/10.1007/978-0-387-30425-0_5.

Lewin, P. A., & Nowicki, A. (1998). Acoustic output levels and ultrasound output display standard. *Archives of Acoustics, 23*(2), 267−280.

Lin, Z., et al. (2020). Non-invasive ultrasonic neuromodulation of neuronal excitability for treatment of epilepsy. *Theranostics, 10*, 5514−5526.

Macé, E., Montaldo, G., Osmanski, B.-F., Cohen, I., Fink, M., & Tanter, M. (2013). Functional ultrasound imaging of the brain: Theory and basic principles. *IEEE Transactions on Ultrasonics, Ferroelectrics, and Frequency Control, 60*, 492−506.

NOOA 2. (2023). https://oceanexplorer.noaa.gov/technology/sonar/side-scan.html. Accessed May 21.

Ophir, J., Cespedes, I., Ponnekanti, H., Yazdi, Y., & Li, X. (1991). Elastography: A quantitative method for imaging the elasticity of biological tissues. *Ultrasonic Imaging, 13*, 111−134.

Ophir, J., Garra, B., Kallel, F., Konofagou, E., Krouskop, T., Righetti, R., et al. (2000). Elastographic imaging. *Ultrasound in Medicine and Biology, 26*, S23−S29.

Ophir, J., Alam, S. K., Garra, B., Kallel, F., Konofagou, E., Krouskop, T., & Varghese, T. (1999). Elastography: Ultrasonic estimation and imaging of the elastic properties of tissues. *Proceedings of the Institute of Mechanical Engineers H, 213*, 203−233.

Pajek, D., & Kullervo Hynynen, K. (2012). Applications of transcranial focused ultrasound surgery. *Acoustics Today, 8*, 8−14.

Palmeri, M. L., & Nightingale, K. R. (2011). Acoustic radiation force-based elasticity imaging methods. *Interface Focus, 1*, 553−564.

Parker, K. J., Doyley, M. M., & Rubens, D. J. (2011). Imaging the elastic properties of tissue: The 20 year perspective. *Physics in Medicine and Biology, 56*, R1–R29.

Powers, J., Porter, T. R., Wilson, S., Averkiou, M., Skyba, D., & Bruce, M. (2000). Ultrasound contrast imaging research. *Medica Mundi, 44*, 28–36.

Qin, S., Caskey, C. F., & Ferrara, K. W. (2009). Ultrasound contrast microbubbles in imaging and therapy: Physical principles and engineering. *Physics in Medicine and Biology, 54*, R27–R57.

Sarvazyan, A., Rudenko, O., Swanson, S., Fowlkes, J., & Emelianov, S. (1998). Shear wave elasticity imaging: A new ultrasonic technology of medical diagnostics. *Ultrasound in Medicine and Biology, 24*, 1419–1435.

Sarvazyan, A. P., Urban, M. W., & Greenleaf, J. F. (2013). Acoustic waves in medical imaging and diagnostics. *Ultrasound in Medicine and Biology, 39*, 1133–1146.

Sarvazyan, A. P., Hall, T. J., Urban, M. W., Fatemi, M., Aglyamov, S. R., & Garra, B. (2011). An overview of elastography: An emerging branch of medical imaging. *Current Medical Imaging Review, 4*(7), 255–282.

Simpson, D. H., Chin, C. T., & Burns, P. N. (1999). Pulse inversion Doppler: A new method for detecting nonlinear echoes from microbubble contrast agents. *IEEE Transactions on Ultrasonics, Ferroelectrics, and Frequency Control, 46*, 372–382.

Song, P., Rubin, J. M., & Lowerison, M. R. (2023). Super-resolution ultrasound microvascular imaging: Is it ready for clinical use? *Zeitschrift fur Medizinische Physik*.

Szabo, T. L. (2011). Chapter on medical imaging. In J. Enderle, S. Blanchard, & J. Bronzino (Eds.), *Introduction to biomedical engineering* (3rd ed). Elsevier Science.

Szabo, T. L. (2021). Hewlett Packard — innovations that transformed diagnostic ultrasound imaging. *Medical Physics International Journal, 6*, 602–620, Special Issue, History of Medical Physics.

Szabo, T. L. (2014a). *Sections 1.3–1.6. Diagnostic ultrasound imaging: Inside out.* (2nd ed.). Elsevier.

Szabo, T. L. (2014b). Imaging systems and applications. *Diagnostic ultrasound imaging: Inside out.* (2nd ed.). Elsevier.

Szabo, T. L. (2014c). Nonlinear acoustics and imaging. *Diagnostic ultrasound imaging: Inside out.* (2nd ed.). Elsevier.

Szabo, T. L. (2014d). Properties of tissue, Appendix C. *Diagnostic ultrasound imaging: Inside out.* (2nd ed.). Elsevier.

Szabo, T. L. (2014e). Ultrasound contrast agents. *Diagnostic ultrasound imaging: Inside out.* (2nd ed.). Elsevier.

Szabo, T. L. (2014f). Elastography. *Diagnostic ultrasound imaging: Inside out.* (2nd ed.). Elsevier.

Szabo, T. L. (2014g). Section 10.11.6. *Diagnostic ultrasound imaging: Inside out.* (2nd ed.). Elsevier.

Szabo, T. L. (2014h). Section 10.12.1. *Diagnostic ultrasound imaging: Inside out.* (2nd ed.). Elsevier.

Szabo, T. L. (2014i). Section 7.9.2. *Diagnostic ultrasound imaging: Inside out.* (2nd ed.). Elsevier.

Szabo, T. L. (2014j). Section 17.2.2. *Diagnostic ultrasound imaging: Inside out.* (2nd ed.). Elsevier.

Szabo, T. L. (2014k). Section 9.7.3. *Diagnostic ultrasound imaging: Inside out.* (2nd ed.). Elsevier.

Szabo, T. L. (2014l). Section 17.3. *Diagnostic ultrasound imaging: Inside out.* (2nd ed). Elsevier.

Tang, J., Postnov, D. D., Kilic, K., Erdener, S. E., Lee, B., Giblin, J. T., Szabo, T. L., & Boas, D. A. (2020). Functional ultrasound speckle decorrelation-based velocimetry of the brain. *Advanced Science, 7*, 2001044.

Tang, M.-X., & Tortoli, P. (2018). Special issue on high frame rate/ultrafast contrast-enhanced ultrasound imaging. *IEEE Transactions on Ultrasonics, Ferroelectrics, and Frequency Control, 65*(12).

Tyler, W. J., Tufail, Y., Finsterwald, M., Tauchmann, M. L., Olson, E. J., & Majestic, C. (2008). Remote excitation of neuronal circuits using low-intensity, low-frequency ultrasound. *PLoS One, 3*, e3511.

Urban, A., Dussaux, C., Martel, G., Brunner, C., Mace, E., & Montaldo, G. (2015). Real-time imaging of brain activity in freely moving rats using functional ultrasound. *Nature Methods, 12*(9), 873–878.

University of Washington, Department of Oceanography Website. (2023). http://ooicruises.ocean.washington.edu/enlighten/file/Hull-mounted + Multibeam + Echosounder.

Verasonics.com. (2023). *Applications gallery.* https://verasonics.com/vantage-application-gallery/#Power-Doppler.

Ward, B., Baker, A. C., & Humphrey, V. F. (1996). Nonlinear propagation applied to the improvement of lateral resolution in medical ultrasound scanners. In Proceedings of 1995 world congress on ultrasonics (pp. 965–968).

Wilson, S. R., & Burns, P. N. (2010). Microbubble-enhanced US in body imaging: what role? *Radiology*, *257*, 24–39.

Woo, J. D. (2023). *Obstetric ultrasound history web*. http://www.ob-ultrasound.net. An excellent web site for the history of medical ultrasound imaging technology and a description of how it works.

Wu, J., & Tong, J. (1997). Measurements of nonlinearity parameter B/A of contrast agents. *Journal of the Acoustical Society of America*, *101*, 1155–1161.

You, Q., Lowerison, M.R., Shin, Y., Chen, X., Sekaran, N.V.C., Dong, Z., Llano, D.A., Anastasio, M. A., Song, P., (2023). Contrast-free super-resolution power doppler (CS-PD) based on deep neural networks. *IEEE Transactions on Ultrasonics Ferroelectrics, and Frequency Control, 70*, 1355–1368.

Zou, J., Meng, L., Lin, Z., Yuan, T., Meng, W., Niu, L., Guo, Y., & Zheng, H. (2020). Ultrasound neuromodulation inhibits seizures in acute epileptic monkeys. *iScience*, *23*, 101066.

Further reading

Acoustical Society of America PACS Index. (n.d.). https://journals.aps.org/PACS.

NOOA 1. (2023). https://oceanexplorer.noaa.gov/explorations/17deepsearch/logs/sept20/sept20.html. Accessed May 21.

NOOA 3. (n.d.). https://www.whoi.edu/what-we-do/explore/instruments/seafloor-mapping-systems/.

Nouhoum, M., Ferrier, J., Osmanski, B. F., Ialy Radio, N., Pezet, S., Tanter, M., & Deffieux, T. (2021). Fully-automatic ultrasound-based neuro-navigation: The functional ultrasound brain GPS. https://doi.org/10.21203/rs.3.rs-382732/v1.

Szabo, T. L. (2014m). Sections 11.11. *Diagnostic ultrasound imaging: Inside out*. (2nd ed.). Elsevier.

Wells, P. N. T., & Liang, H.-D. (2013). Medical ultrasound: Imaging of soft tissue strain and elasticity. *Journal of the Royal Society Interface, 8*, 1521–1549.

Appendix A: Ultrasound resources

Ultrasound presents a large number of words, concepts, and abbreviations and an amazing onslaught of technical information that can be bewildering. For help, we turn to professional societies. Seven major international organizations provide more information, conferences, and journals where the latest ultrasound research is presented and published.

The American Institute of Ultrasound in Medicine (AIUM) meets annually and publishes the latest clinical developments in the *Journal of Ultrasound in Medicine* (JUM) (JUM, 2023). Other resources are available on the AIUM website (AIUM, 2023) and especially useful for those working in medical ultrasound is an ultrasound dictionary (Szabo et al., 2023).

The Acoustical Society of America (ASA) (ASA, 2023) is an international society which meets twice a year and covers a very broad range of topics in acoustics in all media and frequencies. Frequently, presentations and articles on ultrasound topics are published in the *Journal of the Acoustical Society* (JASA) (JASA, 2023). The organization of topics and nomenclature for the JASA can be found in Appendix B.

The main group for ultrasound in the International Institute of Electrical and Electronics Engineers (IEEE) organization is the Ultrasonics, Ferroelectrics, and Frequency Control Society (UFFC-S, 2023). It holds annual meetings called the International Ultrasound Symposium (IUS) and publishes in its journal, *IEEE Transactions on Ultrasonics, Ferroelectrics, and Frequency Control* (T-UFFC) (T-UFFC, 2023). More about the kinds of topics covered by this group (including diverse applications of ultrasound other than medical ultrasound) can be found in Appendix C.

The World Federation of Ultrasound in Medicine and Biology (WFUMB) (WFUMB, 2023) is an organization promoting international use of medical ultrasound, has many resources, and publishes in the *Ultrasound in Medicine and Biology* journal (UMB) (UMB, 2023).

The International Society for Optics and Photonics (SPIE) brings "engineers, scientists, students, and business professionals together to advance light-based science and technology" (SPIE, 2023a). Though primarily directed at optical physics and imaging, the SPIE also holds an annual SPIE Medical Imaging conference in which there is an ultrasound track. Conference proceedings are published yearly (SPIE, 2023b), as is the *Journal of Medical Imaging* (SPIE, 2023c).

The International Society for Therapeutic Ultrasound (ISTU) is dedicated to "bringing knowledge of therapeutic ultrasound to scientific and medical communities around the world," (ISTU, 2023). Since its creation in 2001, "ISTU's principal mission is to foster the diffusion of knowledge concerning the scientific and clinical aspects of therapeutic ultrasound." In addition to therapeutic and surgical ultrasound, topics include neuromodulation and ultrasound imaging dedicated to monitoring therapy progression and assessment of results. The society holds an annual symposium that alternates locations between Europe, Asia, and North America.

Finally, there is the annual International Congress on Ultrasound (ICU) (ICU, 2023) and the affiliated journal, *Ultrasonics* (Ultrasonics, 2023).

References

American Institute of Ultrasound in Medicine. (2023). <https://www.aium.org/> Accessed 10.09.23.

ASA. (2023). <https://acousticalsociety.org/> Accessed 10.09.23.

International Congress on Ultrasound. (2023). <https://www.2021icu.org.cn/> Accessed 10.09.23.

International Society for Therapeutic Ultrasound. (2023). <https://istu.org> Accessed 12.09.23.

Journal of the Acoustical Society. (2023). <https://acousticalsociety.org/asa-publications/> Accessed 10.09.23.

Journal of Ultrasound in Medicine. (2023). <https://onlinelibrary.wiley.com/journal/15509613> Accessed 10.09.23.

SPIE International society for optics and photonics. (2023a). <https://spie.org> Accessed 12.09.23.

SPIE Medical Imaging conference proceedings. (2023b). <https://spic.org/publications/conference-proceedings/browse-by-conference> Accessed 12.09.23.

SPIE Journal of Medical Imaging (JMI). (2023c). <http://bit.ly/44PrgWn> Accessed 12.09.23.

Szabo, T. L., Chen, S., Hoyt, K., & Ziskin, M. C., AIUM terminology subcommittee. (2019). *AIUM Recommended Ultrasound Terminology, Fourth Edition*. <https://bit.ly/48byZkv> Accessed 12.09.23.

Transactions on Ultrasonics, Ferroelectrics, and Frequency Control. (2023). <https://ieee-uffc.org/publication/t-uffc> Accessed 10.09.23.

Ultrasonics. (2023). <https://www.sciencedirect.com/journal/ultrasonics> Accessed 10.09.23.

Ultrasonics, Ferroelectrics, and Frequency Control Society. (2023). <https://ieee-uffc.org/> Accessed 10.09.23.

Ultrasound in Medicine and Biology. (2023). <https://www.umbjournal.org/> Accessed 10.09.23.

World Federation for Ultrasound in Medicine and Biology. (2023). <http://www.wfumb.org> Accessed 10.09.23.

Appendix B: ASA Physical Acoustics Classification Scheme and terminology

Relevant selections from the Acoustical Society of America Physical Acoustics Classification Scheme

Summary of main branches included

43.20.-f General linear acoustics
43.25.-x Nonlinear acoustics
43.30.-k Underwater sound
43.35.-c Ultrasonics, quantum acoustics, and physical effects of sound
43.38.-p Transduction; acoustical devices for the generation and reproduction of sound
43.40.-r Structural acoustics and vibration
43.58.-e Acoustical measurements and instrumentation (see also specific sections for specialized instrumentation)
43.60.-c Acoustic signal processing
43.80.-n Bioacoustics

43.20.-f General linear acoustics

43.20.Bi Mathematical theory of wave propagation (see also 43.40.At)

43.20.Dk Ray acoustics

43.20.El Reflection, refraction, diffraction of acoustic waves (see also 43.30.Es)

43.20.Fn Scattering of acoustic waves (see also 43.30.Ft, Gv, Hw)

43.20.Gp Reflection, refraction, diffraction, interference, and scattering of elastic and poroelastic waves

43.20.Hq Velocity and attenuation of acoustic waves (see also 43.30.Bp, Cq, Es and 43.35.Ae, Bf, Cg)

43.20.Jr Velocity and attenuation of elastic and poroelastic waves

43.20.Ks Standing waves, resonance, normal modes (see also 43.25.Gf, 43.40.At, and 43.55.Br)

43.20.Mv Waveguides, wave propagation in tubes and ducts

43.20.Px Transient radiation and scattering

43.20.Rz Steady-state radiation from sources, impedance, radiation patterns, boundary element methods

43.20.Tb Interaction of vibrating structures with surrounding medium (see also 43.40.Rj)

43.20.Wd Analogies

43.20.Ye Measurement methods and instrumentation (see also 43.58.-e)

43.25.-x Nonlinear acoustics

43.25.Ba Parameters of nonlinearity of the medium

43.25.Cb Macrosonic propagation, finite amplitude sound; shock waves (see also 43.28.Mw and 43.30.Lz)

43.25.Dc Nonlinear acoustics of solids

43.25.Ed Effect of nonlinearity on velocity and attenuation

43.25.Fe Effect of nonlinearity on acoustic surface waves

43.25.Gf Standing waves; resonance (see also 43.20.Ks)

43.25.Hg Interaction of intense sound waves with noise

43.25.Jh Reflection, refraction, interference, scattering, and diffraction of intense sound waves (see also 43.30.Lz and 43.20.Fn)

43.25.Lj Parametric arrays, interaction of sound with sound, virtual sources (see also 43.30.Lz)

43.25.Nm Acoustic streaming

43.25.Qp Radiation pressure (see also 43.58.Pw)

43.25.Rq Solitons, chaos

43.25.Ts Nonlinear acoustical and dynamical systems

43.25.Uv Acoustic levitation

43.25.Vt Intense sound sources

43.25.Yw Nonlinear acoustics of bubbly liquids

43.25.Zx Measurement methods and instrumentation for nonlinear acoustics (see also 43.58.-e)

43.30.-k Underwater sound

43.30.Lz Underwater applications of nonlinear acoustics; explosions (see also 43.25.Cb, Lj)

43.30.Es Velocity, attenuation, refraction, and diffraction in water, Doppler effect

43.30.Ft Volume scattering

43.30.Gv Backscattering, echoes, and reverberation in water due to combinations of boundaries

43.30.Hw Rough interface scattering

43.30.Jx Radiation from objects vibrating under water, acoustic and mechanical impedance (see also 43.58.Bh)

43.30.Pc Ocean parameter estimation by acoustical methods; remote sensing; imaging, inversion, acoustic tomography

43.30.Vh Active sonar systems

43.30.Wi Passive sonar systems and algorithms, matched field processing in underwater acoustics (see also 43.60.Kx)

43.30.Xm Underwater measurement and calibration instrumentation and procedures (see also 43.58.-e)

43.30.Yj Transducers and transducer arrays for underwater sound; transducer calibration (see also 43.58.Vb)

43.30.Zk Experimental modeling

43.35.-c Ultrasonics, quantum acoustics, and physical effects of sound

43.35.Ae Ultrasonic velocity, dispersion, scattering, diffraction, and attenuation in gases

43.35.Bf Ultrasonic velocity, dispersion, scattering, diffraction, and attenuation in liquids, liquid crystals, suspensions, and emulsions (see also 43.30.Es, Ft, Gv, Hw)

43.35.-c Ultrasonics, quantum acoustics, and physical effects of sound

43.35.Cg Ultrasonic velocity, dispersion, scattering, diffraction, and attenuation in solids; elastic constants (see also 43.20.Gp, Jr)

43.35.Dh Pretersonics (sound of frequency above 10 GHz); Brillouin scattering

43.35.Ei Acoustic cavitation in liquids (see also 43.30.Nb)

43.35.Fj Ultrasonic relaxation processes in gases, liquids, and solids

43.35.Gk Phonons in crystal lattices, quantum acoustics (in PACS, see also 63.20.-e)

43.35.Hl Sonoluminescence

43.35.Mr Acoustics of viscoelastic materials

43.35.Vz Chemical effects of ultrasound

43.35.Wa Biological effects of ultrasound, ultrasonic tomography (see also 43.40.Ng and 43.80.Gx, Jz, Sh)

43.35.Xd Nuclear acoustical resonance, acoustical magnetic resonance

43.35.Yb Ultrasonic instrumentation and measurement techniques (see also 43.58.-e)

43.35.Zc Use of ultrasonics in nondestructive testing, industrial processes, and industrial products

43.38.-p Transduction; acoustical devices for the generation and reproduction of sound

43.38.Ar Transducing principles, materials, and structures: general (see also 43.30.Yj and 43.40.Yq)

43.38.Bs Electrostatic transducers

43.38.Fx Piezoelectric and ferroelectric transducers

43.38.Gy Semiconductor transducers

43.38.Hz Transducer arrays, acoustic interaction effects in arrays (see also 43.30.Yj)

43.38.Zp Acoustooptic and photoacoustic transducers (see also 43.35.Sx)

43.40.-r Structural acoustics and vibration

43.40.Le Techniques for nondestructive evaluation and monitoring, acoustic emission (see also 43.35.Zc)

43.40.Ng Effects of vibration and shock on biological systems, including man (see also 43.35.Wa, 43.50.Qp, and 43.80.-n)

43.40.Ph Seismology and geophysical prospecting; seismographs

43.40.Qi

43.58.–e Acoustical measurements and instrumentation (see also specific sections for specialized instrumentation)

43.58.Ls Acoustical lenses and microscopes (see also 43.35.Sx)

43.58.Ry Distortion: frequency, nonlinear, phase, and transient; measurement of distortion

43.58.Ta Computers and computer programs in acoustics (see also 43.75.Wx, 43.55. Ka, 43.60.Gk, and 43.70.Jt)

43.58.Vb Calibration of acoustical devices and systems

43.60.–c Acoustic signal processing

43.60.Ac Theory of acoustic signal processing

43.60.Bf Acoustic signal detection and classification, applications to control systems

43.60.Cg Statistical properties of signals and noise

43.60.Dh Signal processing for communications: telephony and telemetry, sound pickup and reproduction, multimedia

43.60.Ek Acoustic signal coding, morphology, and transformation

43.60.Fg Acoustic array systems and processing, beam-forming

43.60.Gk Space-time signal processing, other than matched field processing (see also 43.35.Sx)

43.60.Hj Time-frequency signal processing, wavelets

43.60.Kx Matched field processing (see also 43.30.Wi)

43.60.Lq Acoustic imaging, displays, pattern recognition, feature extraction

43.60.Mn Adaptive processing

43.60.Np Acoustic signal processing techniques for neural nets and learning systems

43.60.Pt Signal processing techniques for acoustic inverse problems

43.60.Qv Signal processing instrumentation, integrated systems, smart transducers, devices and architectures, displays and interfaces for acoustic systems (see also 43.58.–e)

43.60.Rw Remote sensing methods, acoustic tomography

43.60.Sx Acoustic holography

43.60.Tj Wave front reconstruction, acoustic time-reversal, and phase conjugation

43.60.Uv Model-based signal processing

43.60.Vx Acoustic sensing and acquisition

43.60.Wy Non-stationary signal analysis, non-linear systems, and higher order Statistics

43.80.–n Bioacoustics

43.80.Cs Acoustical characteristics of biological media: molecular species, cellular level tissues

43.80.Ev Acoustical measurement methods in biological systems and media

43.80.Gx Mechanisms of action of acoustic energy on biological systems: physical processes, sites of action (in PACS, see also 87.50.Y-)

43.80.Jz Use of acoustic energy (with or without other forms) in studies of structure and function of biological systems (in PACS, see also 87.50.Y-)

43.80.Qf Medical diagnosis with acoustics (in PACS, see also 87.63.D-)

43.80.Sh Medical use of ultrasonics for tissue modification (permanent and temporary) (in PACS, see also 87.50.Y-)

43.80.Vj Acoustical medical instrumentation and measurement techniques (see also 43.66.Ts and 43.58.-e)

Appendix C: IEEE Ultrasonics, Ferroelectrics, and Frequency Control Society terminology

In the *IEEE Ultrasonics, Ferroelectrics, and Frequency Control Society (UFFC)*, the Ultrasonics group has five subgroups. Many papers are published annually in the *UFFC* transactions. Here are typical topics that fall within the interests of these groups which have been covered in past annual *International Ultrasonics Symposia* and are taken from the 2023 International Ultrasonics Symposium program.

Group 1: Medical Ultrasonics is by far the largest and will be discussed later. Other topics of medical and other interest spill into the other major subgroups of the Ultrasonics Group. Symbol: **M**.

Group 2: Sensors, Nondestructive Evaluation (NDE), and Industrial Applications: acoustic imaging microscopy, general NDE methods, material and defect characterization, structural health monitoring, underwater acoustics, flow measurement, transducer energy harvesting. Symbol: **N**.

Group 3: Physical Acoustics: general physical acoustics, nonlinear acoustics, acoustic tweezers, modeling, optoacoustics, thin films. Symbol: **P**.

Group 4: Microacoustics, surface acoustic wave (SAW), the thin-film bulk acoustic resonator (FBAR), and microelectromechanical systems (MEMS): SAW modeling, MEMS and sensors, BAW applications, materials. Symbol: **A**.

Group 5: Transducer and Transducer Materials: materials and design and fabrication of large arrays, flexible and high frequency arrays, capacitive micromachined ultrasound transducer (CMUT), piezoelectric micromachined ultrasound transducer (PMUT), integrated electronics for ultrasound imaging, transducers for neurostimulation and drug delivery, therapeutic transducers, multiwave and multimode transducers. Symbol: **T**.

Categories: (Note: The labels for these categories begin with the group symbol as the first letter.)

ABD: Bulk acoustic wave devices (BAW)

AMA: Materials for acoustic wave devices

AMD: Microacoustic novel devices

AMR: Microacoustic resonators

ASD: SAW (surface acoustic wave) devices

ASM: SAW modeling

ASS: Sensors and delay lines

MBB: Medical beamforming and beam steering (core beamforming, novel beamforming approaches, sidelobe and clutter mitigation, speed of sound estimation and correction, adaptive and 3D beamforming, artificial intelligence and beamforming tools, coherent transmit compounding)

MBE: Biological effects and dosimetry (stimulatory and therapeutic bioeffects, LIPUS [low intensity pulsed ultrasound], and neuro, mechanisms, and dosimetry)

MBF: Blood flow measurement (cardiovascular and cerebrovascular flow imaging, functional ultrasound and contrast-free microvascular imaging, new blood flow imaging techniques, vector flow imaging, 3D blood flow imaging, deep learning and other new techniques in blood flow imaging, emerging blood flow imaging techniques and applications)

MCA: Contrast agents (bubble technology, contrast agent imaging and quantification, therapy and drug delivery, bubble dynamics, clinical applications and imaging technology, PAM [passive acoustic monitoring], and microbubble production, therapeutic and drug delivery strategies)

MEL: Elastography (Advances in elastography analysis, cardiovascular elastography, elastography of liver and prostate, new elastography methods, wave propagation and elastography in soft tissues, advances in elastography methods, vascular elastography, signal processing for elastography, viscoelasticity)

MIM: Medical imaging (3D/4D imaging, cardiovascular imaging, multi-modal/multi-modality imaging, novel applications of ultrasound imaging, novel imaging techniques, ultrafast Doppler/fUS (functional ultrasound) imaging, advanced acquisition & signal processing techniques, image analysis & AI [artificial intelligence], imaging systems & tomography, US [ultrasound] simulations and US-based modeling)

MIS: Medical image and signal processing (aberration correction, cardiovascular imaging, deep learning for image enhancement, deep learning for image segmentation, image enhancement, imaging, machine learning for image analysis, microvascular and flow imaging, cardiac and blood flow imaging, image classification, image formation and reconstruction, image segmentation, lung ultrasound, motion estimation and image registration, tomography and speed of sound, tracking)

MPA: Medical photoacoustics (functional and molecular imaging, image formation and processing, imaging therapy, system development, clinical applications of photoacoustic imaging, photoacoustic image processing, photoacoustic imaging systems)

MSD: System and device design: (systems for imaging and therapy delivery, systems for imaging and therapy monitoring, ultrasound array systems and applications, wearable systems and devices, medical devices and applications, medical systems and devices)

MSR: Superresolution: (superresolution from the neck down, superresolution on cancer, superresolution on the brain, superresolution techniques, cardiovascular and abdominal superresolution imaging, deep learning for superresolution imaging, signal processing for superresolution imaging, superresolution beamforming and postprocessing, superresolution imaging methods and applications, superresolution imaging of the brain)

MTC: Medical tissue characterization (blood and cardiac tissue characterization, cancer tissue characterization, liver tissue characterization, tissue characterization methods and applications, ultrasound estimation of sound speed, tissue characterization applications, tissue characterization methods)

MTH: Therapeutics, hyperthermia and surgery (cavitation based therapy, drug delivery, neuromodulation, therapy devices, therapy applications)

MTN: Theranostics and image guidance in therapy (image guided theranostics, novel theranostic agents, theranostic treatment monitoring, image guidance, theranostic agents, treatment monitoring)

NAF: Acoustic microfluidics

NUA: Underwater acoustics

NAI: Acoustic imaging and microscopy

NAS: Acoustic sensors

NDE: General NDE methods (acoustic imaging and microscopy; acoustic sensors; structural health monitoring; acoustic microfluidics; photoacoustics; wave propagation, signal processing; material & defect characterization; transducers; process control and industrial ultrasound)

NMC: Material & defect characterization

NPA: Photoacoustics

NWP: Wave propagation

NEH: Energy harvesting

NPC: Process control

NFM: Industrial ultrasound, and flow measurement

NSH: Structural health monitoring

NSP: Signal processing

NTC: Transducers for NDE and industrial applications

PAT: Acoustic tweezers and particle manipulation

PGP: General physical acoustics

PMI: Modeling and inversion

PNL: Nonlinear physical acoustics

POA: Laser ultrasonics and acousto-optics

PPN: Phononics

PTE: High power and temperature effects

PTF: Thin films

PUM: Ultrasonic motors & actuators

TMI: Diagnostic and therapeutic transducers (miniaturized imaging and therapeutic transducers, wearable and flexible transducers, biomedical transducers)

TMS-P: Transducer modeling and electronics

TMU: Micromachined ultrasonic transducers (Capacitive micromachined ultrasonic transducers [CMUT], piezoelectric micromachined ultrasonic transducers [PMUT], optomechanical and electrostrictive transducers, airborne ultrasound transducers)

TPF: Haptics and gesture recognition systems

TPM: Piezoelectric transducer materials and applications

TTT: Miniaturized therapeutic and interventional ultrasound transducers

Appendix D: Tables of material properties

Tables D.1, D.2 and D.3.

Table D.1 Material properties of selected tissues and water.

Material	c_L	α_L	y	Z_L	Source
Blood	1584	0.14	1.21	1.679	1
Bone	3198	3.54	0.9	6.364	1
Brain	1562	0.58	1.3	1.617	1
Breast	1510	0.75	1.5	1.540	1
Fat	1430	0.6	1.0	1.327	1
Heart	1554	0.52	1.0	1.647	1
Kidney	1560	0.24	1.02	1.638	1[a]
Liver	1578	0.45	1.05	1.657	1
Muscle	1580	0.57	1.0	1.645	1
Spleen	1567	0.4	1.3	1.652	1
Water @ 20 °C	1482.3	2.17e-3	2.0	1.482	1

[a]Corrected from the values provided in source 1.

Table D.2 Material properties of selected solid materials.

Material	c_L	α_L	y	α_{L0}	α_{L2}	y_{L2}	Z_L	c_S	α_S	y_S	Z_S	Source
Aluminum 6061 annealed	6350	0.144	1				17.34					2,3
Brass (70Cu,30% Zn) 260 N41	4700	0.045	2				40.6	2100			18.1	2,3
Concrete (Mortar C1)	3710	2.52	1		0.062	4	9.65					2,3
Copper	5010						44.6	2270			20.3	2
Glass (silica) C10	5900	0.058	1				13					2,3
Gold	3240						63.8	1200			23.6	2
Granite C33	6500	11.29	1				17.6					2,3
Graphite	2500	3.84	1		13.9	4	4					2,3
Iron F1	5900	1.29	1				46.4	3200			24.6	2,3
Lexan	2195	4.783	1.001				2.76	943	41.84	0.695	1.13	2,6
Phantom rubber (20°C)	1466	0.19	1.83				1.32					9
Plexiglass	2753	0.941	1	1.29			3.26					2,5
Polyethylene (h.d.)	2566	7.577	0.796				2.463	1.273	26.64	0.815	1.222	2,4,6
Polyethylene (l.d.)	2380	1.522	1.171				2.19	987	22.8	0.95	0.908	2,4,6

(*Continued*)

301

Table D.2 (Continued)

Material	c_L	α_L	y	α_{L0}	α_{L2}	y_{L2}	Z_L	c_S	α_S	y_S	Z_S	Source
Sand (Yellow Sea)	1706	0.0036	1.95				3.07					8
Silcone rubber RTV-560	961.1	3.38	1.428							1.5		10
Silver	3600						38	1600			17	?
Mild steel (Low C) F16	5900	0.225	1				46	3200			25	2,3
Tofu (extra firm)	1.521	0.7055	1.191				1.43					7
Teflon O4	1310	14.71	1				2.87					2,3
Zinc N92	4200	1.58	1		0.11	4	30	2400	29.6		16.8	2,3

Note in some cases, properties from different sources were matched as closely as possible to a material with similar properties.

Attenuation formulae:

Longitudinal: $\alpha\,(dB/cm) = \alpha_{L0} + \alpha_L |f|^y + \alpha_{L2}|f|^{y_{L2}}$.

Shear: $\alpha_S\,(dB/cm) = \alpha_s |f|^{y_s}$.

Table D.3 Material acoustic parameters and their units.

Parameter	Meaning	Units
c_L	Speed of sound (Longitudinal)	m/s
α_L	Attenuation coefficient (Longitudinal, 1st order)	dB/(MHz^y · cm)
y	Exponent (Atten. power law)	–none–
α_{L0}	Attenuation coefficient (Longitudinal, 0th order)	dB/cm
α_{L2}	Attenuation coefficient (Longitudinal, 2nd order)	dB/(MHz$^{y_{L2}}$ · cm)
y_{L2}	Exponent (Atten. power law, Longitudinal, 2nd order)	–none–
Z_L	Impedance (Longitudinal)	MRayl
c_S	Speed of sound (Shear)	m/s
α_S	Attenuation coefficient (Shear, 1st order)	dB/(MHz^{y_S} · cm)
y_S	Exponent (Atten. power law, Shear)	–none–
Z_S	Impedance (Shear)	MRayl

Sources

1. Szabo, T. L. (2014). *Appendix B, Diagnostic ultrasound imaging: Inside Out* (2nd ed.). Oxford, UK: Elsevier.
2. Resources: Acoustic Tables, https://www.ondacorp.com/resources/ Accessed 01.09.23.
3. Ono, K. (2020). Comprehensive report on ultrasonic attenuation of engineering materials, including metals, ceramics, polymers, fiber-reinforced composites, wood, and rocks. *Applied Sciences*, *10*(7), 2230–2281.
4. He, P. (2000). Measurement of acoustic dispersion. Using both transmitted and reflected pulses. *Journal of the Acoustical Society of America*, *107*(2), 801–807.
5. He, P. (1999). Experimental verification of models for determining dispersion from attenuation. *IEEE Transactions on Ultrasonics, Ferroelectrics, and Frequency Control*, *46*(3), 706–714.

6. Szabo, T. L., & Wu, J. (2000). A model for longitudinal and shear wave propagation in viscoelastic media. *Journal of the Acoustical Society of America.*, *107*(5), 2437−2467.
7. Wojcik, G., Szabo, T., Mould, J., Carcione, L., & Clougherty, F. (1999). Nonlinear pulse calculations and data in water and a tissue mimic, *Proceedings of the IEEE 1999 Ultrasonics Symposium (IEEE New York)*, 1521−1526.
8. Zhou, J.-X., Zhang, X.-Z., & Knobles, D. P. (2009). Low-frequency geoacoustic model for the effective properties of sandy seabottoms. *Journal of the Acoustical Society of America*, *125*(5), 2847−2866.
9. Browne, J., Ramnarine, K., Watson, A., & Hoskins, P. (2003). Assessment of the acoustic properties of common tissue-mimicking test phantoms. *Ultrasound in Medicine and Biology*, *29*(7), 1053−1060.
10. Wu, J. (1997). Private communication.

Index

Note: Page numbers followed by "*f*" and "*t*" refer to figures and tables, respectively.